Photodynamic Therapy 2.0

Photodynamic Therapy 2.0

Guest Editors

Stefano Bacci
Kyungsu Kang

Basel • Beijing • Wuhan • Barcelona • Belgrade • Novi Sad • Cluj • Manchester

Guest Editors
Stefano Bacci
Department of Biology
University of Florence
Florence
Italy

Kyungsu Kang
Center for Natural Product
Systems Biology
Korea Institute of Science
and Technology
Gangneung
Republic of Korea

Editorial Office
MDPI AG
Grosspeteranlage 5
4052 Basel, Switzerland

This is a reprint of the Special Issue, published open access by the journal *Biomedicines* (ISSN 2227-9059), freely accessible at: https://www.mdpi.com/journal/biomedicines/special_issues/W50HI24NN4.

For citation purposes, cite each article independently as indicated on the article page online and as indicated below:

Lastname, A.A.; Lastname, B.B. Article Title. *Journal Name* **Year**, *Volume Number*, Page Range.

ISBN 978-3-7258-4971-0 (Hbk)
ISBN 978-3-7258-4972-7 (PDF)
https://doi.org/10.3390/books978-3-7258-4972-7

Contents

About the Editors

Stefano Bacci

Stefano Bacci is a Professor of Cytology and Histology and Developmental Biology at the University of Florence and has resided, during his training, at the Schepens Eye Research Institute, Harvard Medical School, Boston (MA), USA. His main research concerns understanding the cellular mechanisms involved during wound healing in relation to the type of treatment used, including in relation to photodynamic therapy. He is the author of more than 100 articles within his research sector, including publications of high scientific impact.

Kyungsu Kang

Kyungsu Kang is a Principal Researcher at the Korea Institute of Science and Technology (KIST), Republic of Korea. He is also a Professor in Natural Product Applied Science at the University of Science and Technology and an Adjunct Professor in the Department of Convergence Medicine at Yonsei University Wonju College of Medicine. He received his B.S. in Applied Biology and Chemistry, and his M.S. and Ph.D. in Agricultural Biotechnology, from Seoul National University, Republic of Korea. The main research focus of Dr. Kang's laboratory is the discovery of bioactive natural products that promote intestinal health and modulate the gut–organ axis. Dr. Kang and his team are particularly interested in the AI- and data-based discovery of bioactive natural products.

biomedicines

MDPI

Editorial

Photodynamic Therapy 2.0

Kyungsu Kang [1,2] and Stefano Bacci [3,*]

1 Center for Natural Product Systems Biology, Gangneung Institute of Natural Products, Korea Institute of Science and Technology, Gangneung 25451, Gangwon-do, Republic of Korea; kskang@kist.re.kr
2 Natural Product Applied Science, KIST School, University of Science and Technology (UST), Gangneung 25451, Gangwon-do, Republic of Korea
3 Research Unit of Histology and Embryology, Department of Biology, University of Florence, Viale Pieraccini 6, 50139 Florence, Italy
* Correspondence: stefano.bacci@unifi.it

In 1903, Von Tappeiner and Jesionek [1] demonstrated the effectiveness of light therapy when combined with a photosensitizer and oxygen—a phenomenon known as "photodynamic action". Photodynamic therapy (PDT) is widely used in the medical field to treat a variety of oncological and non-oncological human disorders; it generates a high amount of reactive oxygen species that destroy oncological tissue, keratosis, and pathogens. In contrast, recent studies have demonstrated that mild PDT also has beneficial effects, such as wound healing [2], stress and pathogen resistance, and extended lifespan in animal models, such as *Caenorhabditis elegans*, under both normal and pathogen-infected conditions [3,4].

This Editorial, published in *Biomedicines*, provides a current and concise overview of "Photodynamic Therapy 2.0" and discusses recent studies pertaining to PDT. We provide a summary of the various contributions to date, emphasizing novel analytical methods and recent advancements in PDT.

Bowen's disease is a form of cutaneous squamous cell carcinoma (cSCC) that has a high risk of progression to an invasive form [5]. Conventional PDT is a primary treatment option, with varying protocols involving photosensitizers, light sources, and combinations of the two. Dermoscopy and re-reflectance confocal microscopy can monitor the therapeutic response. Treatment is generally well tolerated, with mild side effects and good cosmetic outcomes. A periodic follow-up is necessary because of the risk of recurrence and progression to cSCC [6]. As the incidence of the keratinocyte tumor increases, the opportunity for PDT will expand [7].

PDT has shown promise in treating brain tumors, particularly gliomas [8], with improved median survival rates and minimal side effects compared with conventional methods, such as surgery, radiotherapy, and chemotherapy. This promising approach may influence future treatments for this challenging and deadly disease, and further studies may lead to significant advancements in treating this tumor type [9].

Nasopharyngeal carcinoma, a malignancy linked to the Epstein–Barr virus (EBV), is associated with 140,000 deaths annually [10]. This study aimed to determine the efficacy of PDT in modulating the tumor microenvironment during nasopharyngeal carcinoma treatment. The review included a search of various databases and used the Oral Health Assessment Tool to assess bias. A meta-analysis revealed that PDT treatment significantly increased Interleukin (IL)-8, IL-1α, IL-1β, Microtubule Associated Proteins 1A/1B chain 3b (LC3) BI, LC3BII, Matrix Metallo Proteinase (MMP) 2, and MMP 9 levels in nasopharyngeal carcinoma cells, while reducing Nuclear Factor Kappa-Light-Chain-Enhancer of Activated B cells (NF-κB), miR BamHI-A Region Rightward Transcript (BART) 1-5p, BART 16, and BART 17-5p levels. PDT also reduced apoptosis and the viability of nasopharyngeal cancer cells infected with EBV. This treatment also increased latent membrane protein 1 levels compared with the control group. Further preclinical studies are needed to validate these results [11].

Furthermore, Fernández-Guarino et al. compared the efficacy of PDT with methyl aminolaevulinic acid (MAL) and aminolaevulinic acid (ALA) in a long-term study of multiple actinic keratosis (AK). A total of 46 patients were treated, including 24 for MAL and 22 for ALA. The data indicated no significant differences at 12 months, despite ALA exhibiting slightly superior results after three months. For PDT, both ALA and MAL were efficacious at lower levels of red light. Long-term efficacy was demonstrated; however, additional studies are needed to determine the lowest point of red light exposure without losing efficacy [12]. Ulrich et al. reported the results of a phase III clinical trial using red light PDT plus a BF-200 ALA (10% 5-aminolevulinic acid nanoemulsion) gel on AK. After 12 weeks of PDT in 21 patients, complete lesion clearance rates were 90.9% with PDT compared with 18.6% in the vehicle control group; the lesion recurrence rate with PDT was 29%. Most patients (81%) rated their satisfaction with PDT as very good or good compared with the vehicle control group (42.8%) [13].

Katalinic et al. examined the effect of photoactivating 3% hydrogen peroxide with a 445 nm diode laser on dental implants infected with *Staphylococcus aureus* and *Candida albicans* biofilms. They assigned 80 contaminated titanium implants to the following four groups: (1) negative control (no therapy); (2) positive control (0.2% chlorhexidine); (3) 3% hydrogen peroxide; and (4) photoactivated 3% hydrogen peroxide. A significant difference was observed between all groups and the negative control, which indicates that the new antimicrobial treatment warrants further study and development [14].

Moreover, Rostami et al. examined the effectiveness of a combination of shockwave-enhanced emission photoacoustic streaming (SWEEPS) and PDT with indocyanine green (ICG) for eliminating the biofilm of *Enterococcus faecalis* from infected root canals. After sterilizing and infecting the canals with *E. faecalis* for two weeks, 30 standardized single-canal teeth from healthy humans were used. The teeth were divided into the following six groups: control, ICG, ICG + 808 nm laser, ICG + SWEEPS, ICG + 808 nm laser + SWEEPS, and sodium hypochlorite (NaOCl, 5.25%). The number of colony-forming units per milliliter (CFUs/mL) was determined for each group. Despite significantly reduced bacterial levels in the ICG-treated group in conjunction with an 808 nm diode laser and SWEEPS, the results indicated that NaOCl alone was the most effective. When comparing ICG with an 808 nm diode laser to ICG with SWEEPS, a statistically significant difference was observed. Based on the results, SWEEPS may effectively increase the dispersion of the photosensitizer in the root canal space, and when combined with an irrigant, it provides promising outcomes [15].

Pordel et al. studied the effects of antimicrobial PDT using a blue diode laser (BDL) with varying output powers and photosensitizers (riboflavin and curcumin) for the treatment of *Streptococcus mutans* around orthodontic brackets. A total of 36 orthodontic brackets were contaminated with *S. mutans* and randomly assigned to the following 12 groups: control, riboflavin alone, riboflavin + different powers (200, 300, 400, or 500 mW) of BDL, curcumin alone, curcumin + different powers (200, 300, 400, or 500 mW) of BDL, and 0.2% chlorhexidine (CHX, positive control). Orthodontic brackets were exposed to BDL at a power density of 0.4–1.0 W/cm^2 for 30 s, and mean CFUs/mL were measured before and after treatment. The results indicated that CHX and curcumin plus BDL with an output power of 500 mW showed the greatest reduction in *S. mutans* colony numbers, whereas riboflavin alone and riboflavin + BDL had no significant difference in the control group. The study concluded that antimicrobial photodynamic therapy (aPDT) plus curcumin, as a photosensitizer, and BDL successfully suppresses *S. mutans* colonies around stainless steel brackets [16]. Ha et al. examined the effects of aPDT using red light (660 nm) and a new natural photosensitizer, the ethanol extract of *Ligularia fischeri* (LFE), against various pathogenic bacteria. PDT with 20 µg/mL of LFE and red light (120 W/m^2, 15 min) resulted in a marked antimicrobial activity against methicillin-resistant *S. aureus* (MRSA) as well as *S. mutans*, with a log reduction of 4.7 and 4.9 in viable cells of MRSA and *S. mutans*, respectively. The use of aPDT with LFE showed a much stronger effect compared with vancomycin (100 µg/mL) or ampicillin (100 µg/mL) treatment, with a log reduction of 2.0

and 0.7 in viable cells of MRSA and *S. mutans*, respectively. aPDT decreased the MRSA bacterial cell number in *C. elegans* and extended the lifespan of MRSA-infected worms [17].

Sleep quality is linked to glioma-specific outcomes, including survival [18]. The sleep-induced activation of brain drainage (BD) is important for the survival of glioma patients as it suppresses BD. Photobiomodulation (PBM) is an effective technology for stimulating BD and as an add-on therapy for glioma. A study on male Wistar rats by Shirokow et al. revealed that PBM during sleep stimulates BD more strongly compared with when awake. The study also revealed greater effects of PBM on BD stimulation and the immune response against glioma, including increased CD8+ cells, apoptosis activation, and cell proliferation blockage. This new sleep–phototherapy technology may improve the management of brain cancer patients using smart sleep and non-invasive approaches to glioma treatment [19].

Shakhova et al. examined the efficacy of post-operative phototherapy in mice with CT-26 tumors following excision using a cold knife and a laser scalpel. After the operation, PDT with a photosensitizer based on chlorin was administered at wavelengths of either 405 or 660 nm. The laser scalpel demonstrated superior efficacy compared with the cold knife. The use of PDT following cold knife resection resulted in a decrease in the recurrence rate to 70% and 42% at 405 nm and 660 nm wavelengths, respectively. The use of PDT following laser scalpel resection resulted in recurrence rates of 18% and 30%, respectively. Fluorescence confocal imaging demonstrated that the photosensitizer penetrated more deeply in the cold knife scenario, suggesting a greater effect of PDT at deeper levels. Tumor recurrence was observed in the group exposed to low-dose light without PDT, as evidenced by the disparity in recurrence rates between the 405 nm and 660 nm groups. Light exposure that involved irradiation alone resulted in increased rates of recurrence compared with those administered with PDT. Thus, PDT processing is exclusively advisable for cold knife treatment [20].

In the context of this research, there was an opportunity to meet with professionals in the field of PDT. In terms of content, the reviews that have been recommended are clear and the research articles that have been proposed include innovative techniques and treatment targets that will undoubtedly be further developed in the future. Based on the findings of these studies, it may be concluded that the clinical field is experiencing a greater sense of optimism over the elimination of diseases that have persistently plagued humanity.

Author Contributions: Conceptualization, K.K. and S.B.; writing—review and editing, K.K. and S.B.; funding acquisition, K.K. All authors have read and agreed to the published version of the manuscript.

Funding: This work was supported by an intramural research grant from KIST (2E33301).

Acknowledgments: We acknowledge the authors of the articles referred to in this Editorial for their valuable contributions, as well as the referees for their rigorous review.

Conflicts of Interest: The authors declare no conflicts of interest.

References

1. Grandi, V.; Corsi, A.; Pimpinelli, N.; Bacci, S. Cellular Mechanisms in Acute and Chronic Wounds after PDT Therapy: An Update. *Biomedicines* **2022**, *10*, 1624. [CrossRef] [PubMed]
2. Fernández-Guarino, M.; Bacci, S.; Pérez González, L.A.; Bermejo-Martínez, M.; Cecilia-Matilla, A.; Hernández-Bule, M.L. The Role of Physical Therapies in Wound Healing and Assisted Scarring. *Int. J. Mol. Sci.* **2023**, *24*, 7487. [CrossRef] [PubMed]
3. Nguyen, U.T.T.; Youn, E.; Le, T.A.N.; Ha, N.M.; Tran, S.H.; Lee, S.; Cha, J.W.; Park, J.S.; Kwon, H.C.; Kang, K. Photodynamic Treatment Increases the Lifespan and Oxidative Stress Resistance of *Caenorhabditis elegans*. *Free Radic. Biol. Med.* **2024**, *221*, 98–110. [CrossRef] [PubMed]
4. Muthubharathi, B.C.; Subalakshmi, P.K.; Mounish, B.S.C.; Rao, T.S.; Balamurugan, K. Impact of Low-dose UV-A in *Caenorhabditis elegans* during Candidate Bacterial Infections. *Photochem. Photobiol.* **2024**. *early view*. [CrossRef]
5. Palaniappan, V.; Karthikeyan, K. Bowen's Disease. *Indian Dermatol. Online J.* **2022**, *13*, 177–189. [CrossRef] [PubMed]
6. Arlette, J.P.; Trotter, M.J. Squamous Cell Carcinoma In Situ of the Skin: History, Presentation, Biology and Treatment. *Australas. J. Dermatol.* **2004**, *45*, 1–9. [CrossRef] [PubMed]

7. Antonetti, P.; Pellegrini, C.; Caponio, C.; Bruni, M.; Dragone, L.; Mastrangelo, M.; Esposito, M.; Fargnoli, M.C. Photodynamic Therapy for the Treatment of Bowen's Disease: A Review on Efficacy, Non-invasive Treatment Monitoring, Tolerability, and Cosmetic Outcome. *Biomedicines* **2024**, *12*, 795. [CrossRef] [PubMed]
8. Perry, A.; Wesseling, P. Histologic Classification of Gliomas. *Handb. Clin. Neurol.* **2016**, *134*, 71–95. [CrossRef] [PubMed]
9. Aebisher, D.; Przygórzewska, A.; Myśliwiec, A.; Dynarowicz, K.; Krupka-Olek, M.; Bożek, A.; Kawczyk-Krupka, A.; Bartusik-Aebisher, D. Current Photodynamic Therapy for Glioma Treatment: An Update. *Biomedicines* **2024**, *12*, 375. [CrossRef] [PubMed]
10. Chen, Y.P.; Chan, A.T.C.; Le, Q.T.; Blanchard, P.; Sun, Y.; Ma, J. Nasopharyngeal Carcinoma. *Lancet* **2019**, *394*, 64–80. [CrossRef] [PubMed]
11. Fornel, D.G.; Ferrisse, T.M.; de Oliveira, A.B.; Fontana, C.R. Photodynamic Therapy Can Modulate the Nasopharyngeal Carcinoma Microenvironment Infected with the Epstein–Barr Virus: A Systematic Review and Meta-analysis. *Biomedicines* **2023**, *11*, 1344. [CrossRef] [PubMed]
12. Fernández-Guarino, M.; Fernández-Nieto, D.; Montes, L.V.; Lobo, D.D.P. Methyl Aminolaevulinic Acid versus Aminolaevulinic Acid Photodynamic Therapy of Actinic Keratosis with Low Doses of Red-light LED Illumination: Results of Long-term Follow-up. *Biomedicines* **2022**, *10*, 3218. [CrossRef] [PubMed]
13. Ulrich, M.; Reinhold, U.; Dominicus, R.; Aschoff, R.; Szeimies, R.M.; Schäning, R.; Zeuner, M.T.; Pospiech, N.; Dirschka, T. Effective Treatment of Actinic Keratosis on the Hands with Red Light Photodynamic Therapy Using BF-200 ALA. *Photodiagnosis Photodyn. Ther.* **2024**, *49*, 104280. [CrossRef] [PubMed]
14. Katalinić, I.; Smojver, I.; Morelato, L.; Vuletić, M.; Budimir, A.; Gabrić, D. Evaluation of the Photoactivation Effect of 3% Hydrogen Peroxide in the Disinfection of Dental Implants: In Vitro Study. *Biomedicines* **2023**, *11*, 1002. [CrossRef] [PubMed]
15. Rostami, G.; Afrasiabi, S.; Benedicenti, S.; Signore, A.; Chiniforush, N. The Evaluation of SWEEPS Plus Antimicrobial Photodynamic Therapy with Indocyanine Green in Eliminating Enterococcus faecalis Biofilm from Infected Root Canals: An In Vitro Study. *Biomedicines* **2023**, *11*, 1850. [CrossRef] [PubMed]
16. Pordel, E.; Ghasemi, T.; Afrasiabi, S.; Benedicenti, S.; Signore, A.; Chiniforush, N. The Effect of Different Output Powers of Blue Diode Laser along with Curcumin and Riboflavin against Streptococcus mutans around Orthodontic Brackets: An In Vitro Study. *Biomedicines* **2023**, *11*, 2248. [CrossRef] [PubMed]
17. Ha, N.M.; Hwang, H.; Alam, S.T.; Nguyen, U.T.T.; Lee, S.K.; Park, J.S.; Kim, J.C.; Kwon, H.C.; Kwon, J.; Kang, K. Antimicrobial Photodynamic Therapy with Ligularia fischeri against Methicillin-resistant Staphylococcus aureus Infection in *Caenorhabditis elegans* Model. *Appl. Biol. Chem.* **2023**, *66*, 19. [CrossRef]
18. Allgood, J.E.; Roe, A.; Sparks, B.B.; Castillo, M.; Cruz, A.; Brooks, A.E.; Brooks, B.D. The Correlation of Sleep Disturbance and Location of Glioma Tumors: A Narrative Review. *J. Clin. Med.* **2023**, *12*, 4058. [CrossRef] [PubMed]
19. Shirokov, A.; Blokhina, I.; Fedosov, I.; Ilyukov, E.; Terskov, A.; Myagkov, D.; Tuktarov, D.; Tzoy, M.; Adushkina, V.; Zlatogosrkaya, D.; et al. Different Effects of Phototherapy for Rat Glioma during Sleep and Wakefulness. *Biomedicines* **2024**, *12*, 262. [CrossRef] [PubMed]
20. Shakhova, M.; Elagin, V.; Plekhanov, A.; Khilov, A.; Kurakina, D.; Kamensky, V.; Kirillin, M. Post-operational Photodynamic Therapy of the Tumor Bed: Comparative Analysis for Cold Knife and Laser Scalpel Resection. *Biomedicines* **2024**, *12*, 291. [CrossRef] [PubMed]

biomedicines

MDPI

Brief Report

Methyl Aminolaevulinic Acid versus Aminolaevulinic Acid Photodynamic Therapy of Actinic Keratosis with Low Doses of Red-Light LED Illumination: Results of Long-Term Follow-Up

Montserrat Fernández Guarino *, Diego Fernández-Nieto, Laura Vila Montes and Dario de Perosanz Lobo

Hospital Universitario Ramón y Cajal, Instituto de Investigación Sanitaria Irycis, 28034 Madrid, Spain
* Correspondence: montserrat.fernandezg@salud.madrid.org

Abstract: Photodynamic therapy (PDT) treatment for multiple actinic keratosis (AK) has been found effective when lower doses of red light were used with methyl aminolaevulinic acid (MAL). The aim of this study was to compare the results of lower doses of red light conventional PDT (h-PDT, 16 J/cm^2) with MAL and aminolaevulinic acid (ALA) in a long-term follow-up. Patients with more than five symmetrical AK on the scalp who were candidates for PDT were selected and divided randomly between MAL and ALA treatment and patients were followed at 3 and 12 months. The responses were assessed by counting the total AK and the AK per patient. Pain and adverse events were also compiled. A total of 46 patients were treated, 24 with MAL, and 22 with ALA. The two groups were comparable at baseline ($p > 0.005$). No significant differences were found in the results of both treatments at 12 months, despite ALA exhibiting slightly better results at 3 months. No differences in pain and adverse events were assessed. Both ALA and MAL were effective when lower doses of red light were used in c-PDT. Long term efficacy was also documented. Further studies are necessary to determine the inferior point of red-light illumination without losing efficacy.

Keywords: photodynamic therapy; red light; short illumination

1. Introduction

Photodynamic therapy (PDT) is a non-surgical treatment for non-melanoma skin cancer, indicated in basal cell carcinoma and actinic keratossis (AK). PDT consist in the use of a photosensitizer to be selectively absorbed for the tumoural and premalignant cells, and afterwards destroy these cells with a convenience light source. The photosensitizers most widely used in cermatology are topical, which induce endogenous production of Protoporhirin IX. PPIX is activated by visible light and produces intracellular biological reactions in the tumoral cells via oxygen singlet production (ROS) and necrosis, leading to cellular death [1]. Throughout the decades, PDT has been used with different photosensitizers and light sources. Nowadays, conventional PDT in dermatology is known as the application of a topical photosensitizer, mostly aminolaevulinic acid (ALA) and metylaminolaevulinic acid (MAL), illuminated a red light LED lamp (680 nm, 37 J/cm^2) [2].

Conventional photodynamic therapy (c-PDT) and daylight photodynamic therapy (DL-PDT) have been demonstrated to be effective and comparable treatments for multiple actinic keratosis (AK) [3]. Nevertheless, the difference in the doses of red light used between both modalities, which range from 37 J/cm^2 to a lower total doses of red light in the visible light used for DL-PDT, suggests that maybe a lower dose of red light could be effective in c-PDT. The reason for exploring different forms of illumination in PDT are to relieve pain during the treatment without losing results. Red light-emitting diodes (LED) have shown superiority to other light sources, are the most used devices for performing PDT and are preferred by patients [4]. Optimizing the conventional lamp would be a possible approach to improve tolerance to PDT.

Undoubtedly, DL-PDT has emerged as a great alternative for illumination in PDT, even though there is still a lack of exploration of the influence of the light source and doses used in the global results of PDT. With this argument in mind, we performed a previous study comparing red light conventional illumination (Aktilite®, Galderma, Spain, 630 nm, 37 J/cm^2) with half time illumination with MAL, obtaining similar results [5]. To point up, it seems that the optimal red light doses with the minimal patient discomfort need yet to be defined.

Both aminolaevulinic acid (ALA) and methylaminolaevulinic acid (MAL) have been demonstrated to be effective photosensitizers in c-PDT and DL-PDT obtaining similar results [3,4].

We performed a prospective, comparative, and blind study to assess the efficacy, tolerability, and safety of 17 J/cm^2 of red light doses (h-PDT) for multiple actinic keratosis (AK) with aminolaevulinic acid (ALA) and methylaminolaevulinic acid (MAL) with long-term follow-up.

2. Materials and Methods

Patient candidates were selected for treatment if they had PDT with more than five symmetrically distributed AK of grade I or II on the scalp (Appendix A, Figure A1). The research was conducted between September of 2019 and December of 2021. The study was approved by the Ethics Committee of the hospital, and patients all signed informant consent. Patients were divided randomly into 7.8% ALA gel (Ameluz®, Biofrontera, Germany) or 16% MAL cream (Metvix®, Galderma, Spain) treatment. A nurse trained in PDT procedure, but not otherwise involved in the study, performed randomization.

Age, sex, and phototype of the patients were compiled at the basal visit. The number of total AK in the scalp were counted, mapped, drawn, and classified into grades I and II. Photographs of patients were taken, and AKASI was calculated.

Curettage of grade II AK was performed and dressed after the photosensitizer occlusion for three hours. The illumination (Aktilite®, 630 nm) was shortened into half, in time and doses, and 16 J/cm^2 were applied for 4 minutes.

After PDT, patients evaluated the pain suffered in a visual analogue scale (VAS) from 0 to 10 and were instructed to completely avoid sun exposure in the treated areas for the next 48 hours.

A questionnaire was given in the basal visit to be filled out at home 48 hours after PDT. Patients were instructed to subjectively evaluate the adverse effects from 0 to 3: erythema, edema, crusting, and blistering (0: not present; 1: light; 2: moderate; 3: severe).

The next visits were scheduled 3 and 12 months later, in which AK were assessed. Patients were all evaluated by the same blinded dermatologist, who took no part in the treatment procedure. The primary endpoint was the complete clearance of each AK, and new lesions on the treated area were not evaluated at any time during the follow-ups.

AK complete clearance per patient was compared between groups using Student's *t*-test and a 95% confidence interval (CI), assuming the independency between lesions within patients. The basal characteristics pre-treatment were compared between groups using the two-tailed Student's t-test with a significance value of $p < 0.05$. For the statistical analysis of pain (VAS) and adverse effects, the ANOVA two-tailed test for independent data was used with a significance value of $p < 0.05$.

3. Results

A total of 46 patients completed the study, 24 treated with MAL and 22 with ALA (Table 1). The median age of the patients treated was 77.63 and 80.14, respectively, and all of them were men with phototype II (fair skin). The distribution of AK was comparable in both groups ($p > 0.005$). The MAL group presented a medium basal AKASI of 6.51 (SD 1.17), a total of 27.13 AK per patient (15.88 grade I and 10.33 grade II), with a total of 651 lesions, divided into 391 grade I and 260 grade II. On the other hand, the ALA group had a medium AKASI of 6.81 (SD 1.51), with 30.95 AK per subject (19.23 grade I and 12.64 grade II), and a total of 681 lesions, divided into 413 grade I and 278 grade II.

Table 1. Patients and actinic keratosis (AK) characteristics at the baseline.

		MAL (*N* = 24)	ALA (*N* = 22)	*p* Value
Age		77.63 (±8.41)	80.14 (±4.93)	$p = 0.229$
Sex		24M/0F	22M/0F	NA
Phototype		2.13 (±0.34)	2.05 (±2.13)	$p = 0.350$
BASAL AKASI ($AKASI_0$) *		6.51 (±1.17)	6.81 (±1.51)	$p = 0.438$
Total lesions per subject Mean ± SD		27.13 (±10.34)	30.95 (±8.42)	$p = 0.178$
AKP0* grade I		15.88 (±8.59)	19.23 (±6.82)	$p = 0.152$
AKP0* grade II		10.33 (±6.82)	12.64 (±5.82)	$p = 0.316$
Total lesions (*n*)	1332	651	681	
Grade I	804	391	413	$p = 0.239$
Grade II	532	260	278	

* $AKASI_0$: Total AKASI of the sample at baseline expressed in mean and standard deviation; AKP0: Total count of AK per patient at baseline. Comparative groups $p > 0.005$.

The results of the comparison of the efficacy of both treatments are summarized in Table 2. After the treatment, both groups of patients improved, but with statistical differences; the ALA group achieved a better response (7.77 vs. 14.59, $p = 0.016$) at 3 months with a persistence of 25% of the overall AK (vs. 52% with MAL, $p = 002$). At 12 months, no statistical differences were found ($p = 0.22$) with a mean of lesion per patient of 13.80 with MAL (overall 46%) and 8.09 (overall 26%) with ALA.

Table 2. Summary of the comparative results per patient and total lesions.

	MAL (*N* = 24)	ALA (*N* = 22)	*p* Value *
BASAL			
Total basal AK/per subject (AKP0)	27.13 (±10.34)	30.95 (±8.42)	$p = 0.178$
Total AK = 1332	651	681	
3 MONTHS			
Total AK/per subject (AKP3)	14.59 (±11.32)	7.77 (±5.9)	$p = 0.016$
Total AK = 511	341/651 (52%)	171/681 (25%)	$p = 0.02$
12 MONTHS			
Total AK/patients at 12 months (AKP12)	13.80 (±9.15)	8.09 (±4.80)	$p = 0.22$
Total AK = 476	298/651 (46%)	178/681 (26%)	$p = 0.244$

* t: Student independent data.

Pain during PDT was similar in both groups, with a VAS of 5.21 for MAL and 5.31 for ALA (Table 3). With respect to the adverse events, edema and blisters reached zero punctuation, so they were not analysed. The presence of erythema and crusts were similar and low without statistical differences (Table 3). The VAS after treatment was near to none with no differences ($p = 0.221$).

Table 3. Comparison of the pain in a visual analogue scale (VAS, 0–10)) and local side effects (LSE).

	MAL (*N* = 24)	ALA (*N* = 22)	*p* Value *
VAS	5.21 (±2.3)	5.31 (±1.64)	$p = 0.32$
Erythema	1.3 (±0.48)	1.09 (±0.29)	$p = 0.854$
Edema	------	------	NA
Crusts	0.25 (±0.61)	0.41 (±0.66)	$p = 0.345$
Blisters	------	------	NA

* ANOVA two-tailed test independent data.

4. Discussion

The reasons for exploring different light sources in PDT include obtaining extra benefits as intense pulsed light (IPL) or lasers in rejuvenation, shortening the time of illumination [6], decreasing pain [4], or simplifying the technique with LED [3]. A lot of new LED-based devices have proven efficacy [7], but the most practical approach is likely to optimize the most widely used traditional LED lamp.

Since PDT was first implemented, 37 J/cm^2 were used in the c-PDT protocol (Aktilite®), despite the fact that they were painful and produced side effects such as erythema, crusts, and blisters [3,4]. Subsequently, the development of DL-PDT expanded the knowledge of illumination in a moderate way, and lower doses of red light were applied [5], demonstrating the same effectivity with more local damage (ROS production), and suggesting that the determinant factor for cytotoxicity is the total doses delivered, and not the irradiance or the PpIX accumulation [8]. In a comparative study modeling the local damage in PDT, DL-PDT achieved more local damage than C-PDT, suggesting that higher doses of red light are not related with higher lesion destruction [9] Moreover, when different protocols of illuminations are compared, the best option is likely the one with the best results at three months without pain [9], a remarkably practical approach.

Immunosuppression with high red light doses of PDT have been found after treatment, not only local but also systemic. It is worrisome that the capacity for fostering tumours in the treated area could be related [10]. Reaching a correct immune memory response is more beneficial [11]. These arguments are similar to the principles of photobiomodulation, which uses LED-light properties without a photosensitizer in modulating biological effects, and consequently, lower doses with conventional red light LED illumination could be as effective [5], non-suppressive, restorative, and less painful [12]. None of the patients treated in this study developed any malignancy during the follow-up period, nevertheless, it was a small sample. In the literature review, after the application of DL-PDT, no tumours had been described either [4–13].

In a previous study, half-time illumination (h-PDT) with c-PDT showed similar efficacy to c-PDT [5], and similar efficacy was achieved in h-PDT with ALA and MAL. Both groups of patients improved, achieving a maintenance response which was eventually not significant at 12 months. However, patients treated with ALA reached better significant results at 3 months (p = 0.016). The results in both groups, comparable at the baseline (p = 0.178, Table 1), were low in comparison with other studies in c-PDT [3,5,14] (54% with MAL and 74% with ALA overall reduction). Nevertheless, the severity of the patients treated with a mean basal AKASI higher than six (Table 1) and a mean of 27.13 AK per subject in the MAL group and 30.95 in the ALA group, with only one session applied, should be considered [15].

The side effects were mild, erythema being the most frequent with a medium punctuation of 1.3 out of 4 for MAL and 1.09 out of 4 for ALA (Table 3) without significant differences. Other side effects assessed had a very low score. There is no established protocol for evaluating side-effects after PDT [15], and the scale of our study was filled in by patients at home and was not validated with the consequent limitation.

With respect to pain, both photosensitizers exhibited similar punctuation in VAS, 5.21 with MAL and 5.31 with ALA on a scale from 0 to 10. The most frequent scale used to evaluate pain during PDT in the literature is the visual analogue scale (VAS) from 0 to 10, and c-PDT usually appeared with a mean of 4.4 to 5.7 in the VAS, and daylight was nearly painless [4,16,17]. Thus, h-PDT continued to be painful, the difference being that that pain lasts half the time. In a practical sense, if a patient had unbearable pain, the illumination could be shortened without losing efficacy if it had been at least a half.

5. Conclusions

In conclusion, both ALA and MAL are comparable and effective when 16 J/cm^2 of red light in c-PDT is used. This protocol could be used to relieve pain during illumination.

Further studies are necessary to truly assess if local immunosuppression is avoided and what the cost of effectivity is when illumination is shortened.

Author Contributions: Conceptualization, M.F.G. and D.d.P.L.; methodology, M.F.G.; software, M.F.G.; validation, M.F.G., D.F.-N. and L.V.M.; formal analysis, M.F.G.; investigation, M.F.G. and L.V.M.; resources, M.F.G.; data curation, M.F.G.; writing—original draft preparation, M.F.G.; writing—review and editing, M.F.G.; visualization, M.F.G.; supervision, D.d.P.L. All authors have read and agreed to the published version of the manuscript.

Funding: This research received no external funding.

Institutional Review Board Statement: The study was conducted in accordance with the Declaration of Helsinki, and approved by the Ethics Committee of the Hospital Ramón y Cajal (protocol code 032-19 on the 16 April 2019).

Informed Consent Statement: Informed consent was obtained from all subjects involved in the study.

Data Availability Statement: Not applicable.

Conflicts of Interest: The authors declare no conflict of interest.

Appendix A

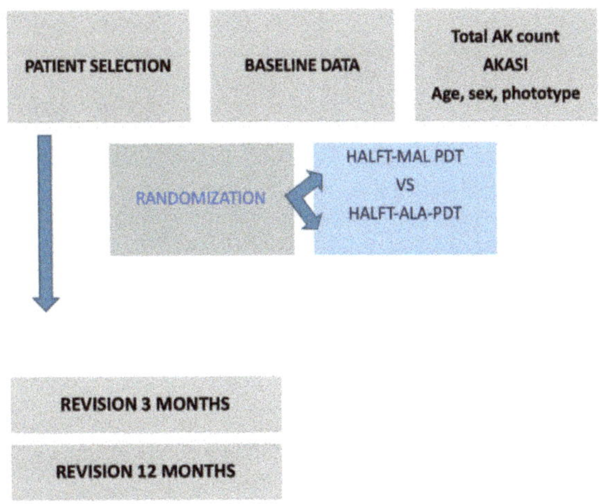

Figure A1. Diagram of the study.

References

1. Kalka, K.; Merk, H.; Mukhtar, H. Photodynamic therapy in dermatology. *J. Am. Acad. Dermatol.* **2000**, *42*, 389–413. [CrossRef] [PubMed]
2. Babilas, P.; Landthaler, M.; Szeimies, R.M. Photodynamic therapy in dermatology. *Eur. J. Dermatol.* **2006**, *16*, 340–348. [PubMed]
3. Morton, C.A.; Szeimies, R.M.; Sidoroff, A.; Braathen, L.R. European guidelines for topical photodynamic therapy part 1: Treatment delivery and current indications-actinic keratoses, Bowen's disease, basal cell carcinoma. *J. Eur. Acad. Derm. Venereol.* **2013**, *27*, 536–544. [CrossRef] [PubMed]
4. Lacour, J.P.; Ulrich, C.; Gilaberte, Y.; Von Felrbert, V.; Basset-Seguin, N.; Dreno, B.; Girard, C.; Redondo, P.; Serra-Guillen, C.; Synnerstad, I.; et al. Daylight photodynamic therapy with methylaminolevulinate cream is effective and nearly painless in treating actinic keratosis: A randomized, investigator-blinded, controlled, phase III study throughout Europe. *J. Eur. Acad. Derm.* **2015**, *29*, 2354–2358.
5. Fernández-Guarino, M.; Fonda Pascual, P.; Lizuain Gomez, P.; Harto Castaño, A.; Jaén Olasolo, P. Split-face study comparing conventional MAL photodynamic therapy in multiple actinic keratosis with complete time vs. half-time red light LED conventional illumination. *J. Eur. Acad. Derm. Venereol.* **2019**, *33*, 1529–1534. [CrossRef] [PubMed]
6. Kessels, J.P.; Nelemans, P.J.; Mosterd, K.; Kelleners- Smeets, N.W.; Krekels, G.A.; Ostertag, J.U. Laser-mediated photodynamic therapy: An alternative Teratment for Actinic Keratosis. *Acta Derm Venereol.* **2016**, *96*, 151–354. [CrossRef] [PubMed]

7. Mordon, S.; Vignion-Dewalle, A.S.; Abi-Rached, H.; Thecua, E.; Lecomte, F.; Vicentini, C.; Deleporte, P.; Béhal, H.; Kerob, D.; Hommel, T.; et al. The conventional protocol vs. a protocol including illumination with a fabric-based biophotonic device (the Phosistos protocol) in photodynamic therapy for actinic keratosis: A randomized, controlled, noninferiority clinical study. *Br. J. Derm.* **2020**, *182*, 76–84. [CrossRef] [PubMed]
8. Nissen, C.V.; Heerfordt, I.M.; Wiegell, S.R.; Mikkelsen, C.S.; Wulf, H.C. Increased protoporphyrin IX accumulation does not improve the effect of photodynamic therapy for actinic keratosis: A randomized controlled trial. *Br. J. Derm.* **2017**, *176*, 1241–1246. [CrossRef] [PubMed]
9. Vignion-Dewalle, A.S.; Baert, G.; Thecua, E.; Vicentini, C.; Mortier, L.; Mordon, S. Photodynamic therapy for actinic keratosis: Is the European consensus protocol for daylight PDT superior to conventional protocol for Aktilite CL 128 PDT? *J. Photochem. Photobiol. B* **2017**, *174*, 70–77. [CrossRef] [PubMed]
10. Vignion-Dewalle, A.S.; Baert, G.; Thecua, E.; Lecomie, F.; Vicentini, C.; Abi-Rached, H.; Mortier, L.; Mordron, S. Comparison of 10 efficient protocols for photodynamic therapy of actinic keratosis: How relevant are effective light dose and local damage in predicting the complete response rate at 3 months? *Lasers Surg. Med.* **2018**, *50*, 576–589. [CrossRef] [PubMed]
11. Anand, S.; Govande, M.; Yasinchak, A.; Heusinkveld, L.; Shakya, S.; Fairchild, R.L.; Maytin, E.V. Painless Photodynamic Therapy Triggers Innate and Adaptive Immune Responses in a Murine Model of UV-induced Squamous Skin Pre-cancer. *Photochem. Photobiol.* **2021**, *97*, 607–617. [CrossRef] [PubMed]
12. Jung, A.C.; Moinard-Butot, F.; Thibaudeau, C.; Gasser, G.; Gaiddon, C. Antitumor Immune Response Triggered by Metal-Based Photosensitizers for Photodynamic Therapy: Where Are We? *Pharmaceutics* **2021**, *13*, 1788. [CrossRef] [PubMed]
13. Avci, P.; Gupta, A.; Sadasivam, M.; Vecchio, D.; Pam, Z.; Pam, N.; Hamblin, M.R. Low-level laser (light) therapy (LLLT) in skin: Stimulating, healing, restoring. *Semin. Cutan. Med. Surg.* **2013**, *32*, 41–52. [PubMed]
14. Fargnoli, M.C.; Piccioni, A.; Neri, L.; Tambone, S.; Pellegrini, C.; Peris, K. Conventional vs. daylight methyl aminolevulinate photodynamic therapy for actinic keratosis of the face and scalp: An intra-patient, prospective, comparison study in Italy. *J. Eur. Acad. Derm. Venereol.* **2015**, *29*, 1926–1932. [CrossRef] [PubMed]
15. García-Rodrigo, C.G.; Pellegrini, C.; Piccioni, A.; Tambone, S.; Fargnoli, M.C. Single versus two-treatment schedule of methyl aminolevulinate daylight photodynamic therapy for actinic keratosis of the face and scalp: An intra-patient randomized trial. *Photodiagnosis Photodyn. Ther.* **2019**, *27*, 100–104. [CrossRef] [PubMed]
16. Gholam, P.; Kroehl, V.; Enk, A.H. Dermatology life quality index and side effects after topical photodynamic therapy of actinic keratosis. *Dermatology* **2013**, *226*, 253–259. [CrossRef] [PubMed]
17. Sotiriou, E.; Evangelou, G.; Papadavid, E.; Apalla, Z.; Vrani, F.; Vakirlis, E.; Panagiotou, M.; Stefanidou, M.; Pombou, T.; Krasagakis, K.; et al. Conventional vs daylight photodynamic therapy for patients with actinic keratosis on face and scalp: 12-months follow-up results of a randomized, intra-individual comparative analysis. *J. Eur. Acad. Dermatol.* **2018**, *32*, 595–600. [CrossRef] [PubMed]

biomedicines

MDPI

Article

Evaluation of the Photoactivation Effect of 3% Hydrogen Peroxide in the Disinfection of Dental Implants: In Vitro Study

Ivan Katalinić [1], Igor Smojver [1], Luka Morelato [2], Marko Vuletić [3], Ana Budimir [4] and Dragana Gabrić [3,*]

1 Specialty Hospital St. Catherine, 10000 Zagreb, Croatia; ikatalinic87@gmail.com (I.K.)
2 Department of Oral Surgery, Faculty of Dental Medicine, University of Rijeka, 51000 Rijeka, Croatia
3 Department of Oral Surgery, School of Dental Medicine, University Hospital Centre Zagreb, University of Zagreb, 10000 Zagreb, Croatia
4 Department of Clinical and Molecular Microbiology, School of Medicine, University Hospital Centre Zagreb, University of Zagreb, 10000 Zagreb, Croatia
* Correspondence: dgabric@sfzg.hr

Abstract: Photoactivation of 3% hydrogen peroxide with a 445 nm diode laser represents a relatively new, insufficiently researched antimicrobial method in the treatment of peri-implantitis. The purpose of this work is to evaluate the effect of photoactivation of 3% hydrogen peroxide with a 445 nm diode laser, and to compare the obtained results with 0.2% chlorhexidine treatment and 3% hydrogen peroxide treatment without photoactivation, in vitro, on the surface of dental implants contaminated with *S. aureus* and *C. albicans* biofilms. Previously, 80 infected titanium implants with *S. aureus* and *C. albicans* cultures were divided into four groups: G1-negative control (no treatment), G2-positive control (0.2% chlorhexidine), G3 (3% hydrogen peroxide), and G4 (photoactivated 3% hydrogen peroxide). The number of viable microbes in each sample was determined by the colony forming unit (CFU) count. The results were statistically processed and analyzed, showing a statistically significant difference across all groups compared to the negative control (G1), and the absence of a statistically significant difference between groups G1–G3. The new antimicrobial treatment, according to the results, could be worthy of further analysis and research.

Keywords: photoactivation; 3% hydrogen peroxide; diode laser; periimplantitis therapy

1. Introduction

Dental implants present a valuable therapeutic choice in the treatment of partial and complete edentulousness. However, despite the high degree of success of dental implants, vast experience and technological improvements during the last 60 years, their frequent use has also brought some very specific problems [1,2]. Peri-implant diseases (peri-implant mucositis and peri-implantitis) are probably the most significant issues associated with dental implants [3,4]. Peri-implant mucositis is defined as the presence of an inflammatory infiltrate in soft tissue induced by plaque, without loss of peri-implant bone, while peri-implantitis is a condition in which, along with soft tissue inflammation, a bone loss is present [5]. If peri-implantitis is not recognized in time and properly treated, implant loss may happen. According to Atieh et al. [6], as many as 50% of placed implants show signs of peri-mucositis, and 12–43% of implants show signs of peri-implantitis. The role of microorganisms in the development of these pathological conditions is crucial and more than 20 different species are routinely found in swabs taken from infected implants [5,7]. One specific microorganism, Staphylococcus aureus, plays a significant role in the development of peri-implantitis due to its affinity for the titanium surface of the implant and the formation of biofilms or a substrate for the growth of biofilms of other cultures, the so-called "early colonizer" [8,9]. Biofilm is defined as a microbial community of cells embedded in a polymer extracellular matrix, the creation of which is caused by microorganisms present in the matrix. Organisms inside biofilms are more resistant to different antimicrobial treatments than free, planktonic organisms [10]. Candida albicans is the most frequently isolated fungus in the human oral cavity. Despite being considered a commensal species, under certain conditions, such as periods of antibiotic use or periods of immunosuppression, it can cause mucosal infection [11]. *C. albicans* is often present in the peri-implant sulcus, both in healthy people and in patients with peri-implantitis, where it also creates a substrate for the formation of biofilms and supports inflammation [12].

The treatment of peri-implant diseases is a complex task due to the difficult access to the implant surface, limited visibility, the roughness of the titanium surface, and the pathogenicity of the microorganisms involved in the pathogenesis of these diseases [13–15]. Therefore, over time, numerous antimicrobial procedures and protocols have been developed. Scarano et al. [14] made an overview of the available treatments in a following manner: (1) mechanical debridement using plastic curettes, rubber polishers, ultrasonic scalers or air-powder abrasives; (2) chemical decontamination using chlorhexidine, citric acid, tetracycline, hydrogen peroxide, etc.; (3) dental laser-based treatments. Described procedures and techniques can also be mixed or combined, as in the newer study carried out by Alovisi et al. [16], where triple antibiotic paste and a glycine powder air-flow abrasion were used to fight microorganisms. However, none of the usual methods can completely remove or inactivate peri-implant pathogens due to previously mentioned factors, such as complex anatomical relationships and/or the specific implant surface [17]. Furthermore, it is even possible to damage this implant surface using certain antimicrobial treatments agents, which can then impair the healing of peri-implant tissues [18].

Although many treatment modalities exist, chlorhexidine digluconate (CHX) combined with manual debridement is still considered to be the golden standard in the treatment of periodontal and peri-implant diseases [19]. Hydrogen peroxide (H_2O_2) has been used in dentistry as a mouthwash to prevent plaque and as an antiseptic after oral surgery for more than 100 years [20], and it is also used in the treatment of peri-implantitis, but likely not as often as CHX. Photo-activation of H_2O_2, with the aim of improving the antimicrobial effect in periodontitis therapy, is a potentially interesting idea examined in the work of Mahdi et al. in 2015 [21].

They found a stronger disinfection potential when H_2O_2 was activated by LED light with a wavelength of 440–480 nm, compared to non-activated H_2O_2. Guided by this idea, the authors of this study wanted to examine the possibility of H_2O_2 activation with a 445 nm diode laser, and augmentation of its antimicrobial effect. According to authors knowledge, there is currently no research on this topic.

The purpose of this work is to evaluate the effect of photoactivation of 3% hydrogen peroxide with a 445 nm diode laser and compare the obtained values with chlorhexidine and hydrogen peroxide without photoactivation, in vitro, on the surface of implants contaminated with *S. aureus* and *C. albicans* biofilms.

The hypotheses of this study are:

1. Hydrogen peroxide activated by a 445 nm diode laser shows better results in disinfection of dental implants compared to hydrogen peroxide used without 445 nm photoactivation.
2. Hydrogen peroxide activated by a 445 nm diode laser shows equal or better results in disinfection of dental implants compared to chlorhexidine treatment.

2. Materials and Methods

The research was conducted at the Department of Oral Surgery of the School of dental medicine in Zagreb and at the Clinical Institute for Clinical and Molecular Microbiology of KBC Zagreb, Croatia. Eighty Zimmer Biomet Tapered Screw-Vent MTX 4.1/10 mm titanium implants (Zimmer Biomet, Palm Beach Gardens, FL, USA) were used in the study. According to Ehrenfest et al. [22], these implants are grade 5 titanium core, with a surface modified by sand-blasting (resorbable blasting media—RBM like calcium phosphate). The surface was microrough, nano-smooth and homogenous. The implants used in this study were contaminated with cultures of *S. aureus* and *C. albicans* isolated from clinical samples at University hospital centre Zagreb. Bacterial and fungal strains were grown separately on Columbia agar for 72 h, after preparation of separate bacterial and fungal suspensions using a thioglycolate broth. They were then mixed into a common suspension. An optical densitometer (Densimat, Biomerieux, Marcyl'Etoile, France) determined the density at 600 nm, which corresponded to 1×10^8 CFU/mL. All dental implants were immersed in 0.3 mL of mixed bacterial–fungal suspension for 14 days under aerobic conditions, at a temperature of 35 °C. The suspension was contained with *S. aureus* and *C. albicans*, at a density of 0.5 McFarland. After that, all implants were randomly distributed into four groups, so that there were 20 implants per group (Figure 1). The implants were removed from the test tubes with sterile tweezers. The implants were then put on sterile gauze and, depending on which group they were in, went through the right cleaning process.

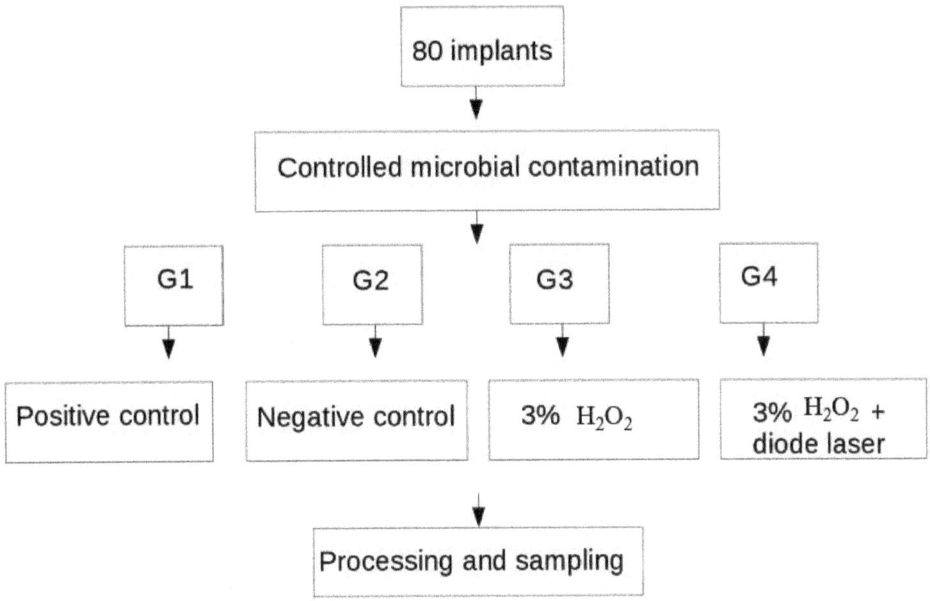

Figure 1. Test sample distribution, contamination and disinfection outline.

2.1. Disinfection Protocols

The first group (G1) represented the negative control. This group was not subjected to any disinfection protocol and served as a reference in assessing the effectiveness of a particular disinfection protocol.

The second group (G2) represented the positive control. That group was treated with 0.2% CHX. CHX was applied on a sterile cotton pellet, and the surface of the implant was rubbed with it for 60 s.

The third group (G3) was treated with 3% hydrogen peroxide. H_2O_2 was applied on a sterile cotton pellet, and the surface of the implant was rubbed with it for 60 s.

The fourth group (G4) underwent the same procedure as G3, with the following difference: 60 s after treatment with 3% hydrogen peroxide, the samples were illuminated with a SiroLaser Blue laser (Dentsply Sirona, Bensheim, Germany) with the following parameters: wavelength of 445 nm, power of 1 W, continuous beam (CW), 320 mm laser fiber tip and 60 s exposure time (Figure 2). The distance of the laser tip from the implant surface was approximately 1–2 mm, with a constant movement of 1 mm/s along the implant surface (freehand movement).

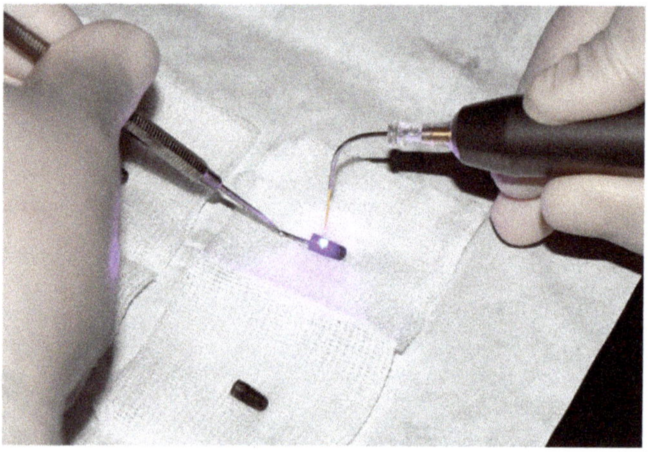

Figure 2. Surface decontamination with SiroLaser Blue laser (Dentsply Sirona, Bensheim, Germany) after treatment with 3% hydrogen peroxide.

2.2. Collection of Samples

After the processing of the implants, the surface of each of them was scraped with the help of a sterile plastic inoculating loop in such a way, that the threads of the implants were scraped with it twice (Figure 3). Each sample was then placed in 250 µL of saline and vortexed on a vibro-mixer (Corning® LSETM vortex mixer, Corning, NY, USA) for 40 s to separate bacteria and fungi. Then, the loops were discarded, and the physiological solution into which the bacteria and fungi were scraped from the implant were transferred with to the blood agar, in a volume of 50 µL. Then, 100 µL of physiological solution, with suspended *S. aureus* and *C. albicans*, was transferred to the wells of a microtiter plate filled with 100 µL of brain–heart broth.

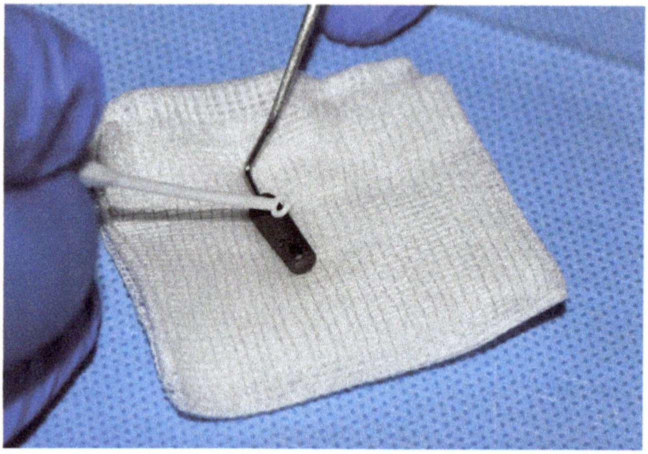

Figure 3. Implant surface sampling with sterile plastic inoculating loop. From the first well, which contains a total of 200 µL (100 µL suspension + 100 µL brain–heart broth), 100 µL was transferred to the next well, and so on, until a dilution of 10^{-7} was reached (Figure 4). From each well, 50 µL was inoculated onto blood agar. The plates were then incubated at 35 °C for 48 h. Then, the growth of microbial cultures was read in such a way, that the CFU of individual pathogens was counted, and the number was multiplied by the degree of the dilution (Figure 5).

The formation of bacterial and fungal plaques was confirmed through an SEM device (JSM-7800 F Schottky Field Emission Scanning Electron Microscope, JEOL Ltd., Tokyo, Japan), as seen in Figure 6.

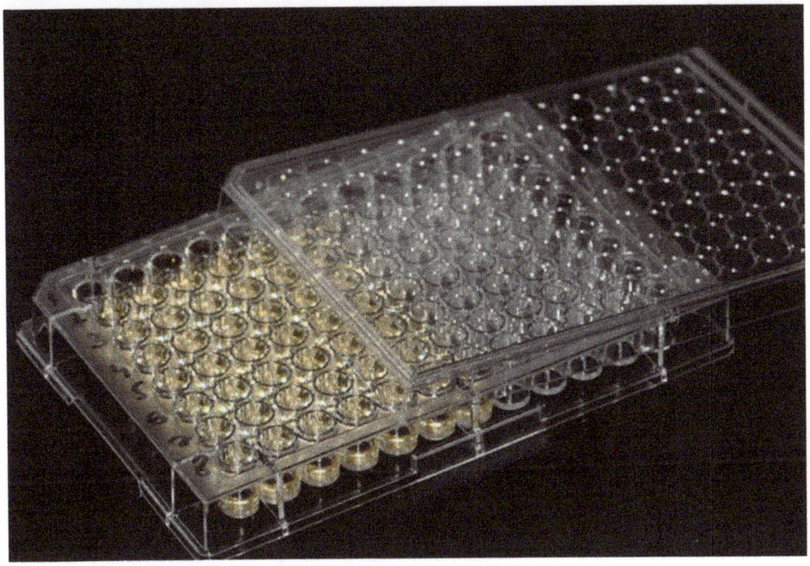

Figure 4. Dilution of samples up to a concentration of 10^{-7}.

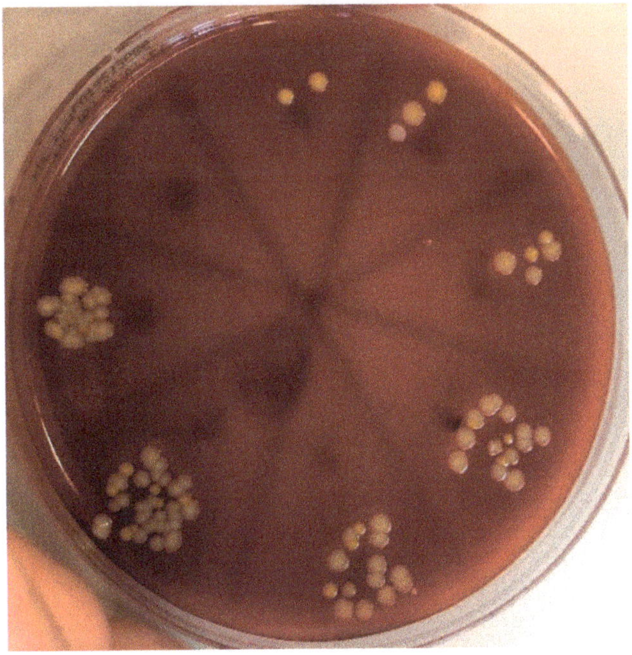

Figure 5. The growth of microbial cultures, ready for CFU/mL counting.

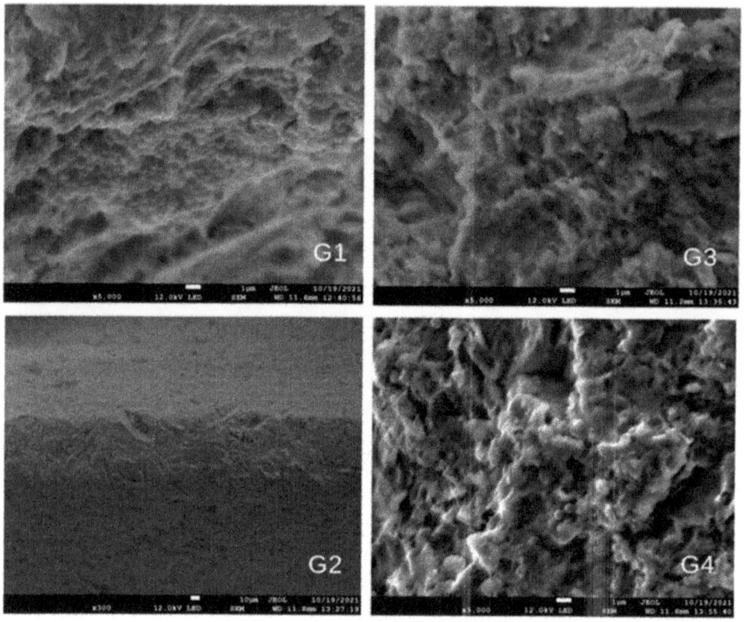

Figure 6. SEM images of G1–G4. Microbial biofilms were visible on the surface of the implant in G1, while they were practically absent in the remaining images.

2.3. Statistical Analysis

Descriptive statistics of CFU/mL were performed for *S. aureus* and *C. albicans*, depending on the disinfection protocol implemented. Furthermore, a Kruskal–Wallis ANOVA by Ranks analysis was performed, especially for *S. aureus* and especially for *C. albicans*, to prove whether there was a difference between the studied groups. It was followed by post hoc multiple comparison *p*-values (2-tailed), in order to show where there is a statistically significant difference. Then, the protocol was compared individually by the Mann–Whitney U test with continuous correction. Statistical calculations were performed with the TIBCO Data Science Workbench (TIBCO Software, Inc., Palo Alto, CA, USA), software version 14.0.0.15.

3. Results

Descriptive statistics of CFU/mL for all protocols for *S. aureus* are shown in Table 1. In the negative control, the highest proportion of colonies per mL was present at 2.86×10^8, followed by the group disinfected with hydrogen peroxide (G3) with 134 CFU/mL, the group of laser-activated hydrogen peroxide (G4) with 45 CFU/mL, and the positive control (G2) with the smallest proportion of 4 CFU/mL. Differences between groups (Mann–Whitney U test) are shown in Table 2 and graphically in Figure 7.

Table 1. Descriptive statistics of the amount of CFU/mL of *S. aureus* under different protocols.

CFU/mL *S. aureus*	Sample Size	Medium Value	Minimal Value	Maximal Value	Standard Deviation
Negative control	20	286,731,516	20.00	2,000,000,000	508,403,118
Positive control	20	4	0.00	20	8.21
3% H_2O_2	20	134	0.00	2000	449.75
Laser + 3% H_2O_2	20	45	0.00	200	79.97

CFU/mL—colony-forming units per milliliter.

Table 2. Comparison of the success of individual protocols in the eradication of *S. aureus*: G1–negative control, G2–positive control, G3–hydrogen peroxide, and G4–laser-activated hydrogen peroxide.

Mann-Whitney U Test By Variable CFU/mL *S. aureus* Marked Tests Are Significant at $p < 0.05$		
Protocol	p-value	2×1 sided exact p
G1 vs. G2	**0.000000**	**0.000000**
G2 vs. G3	0.059363	0.120700
G2 vs. G4	0.326615	0.461169
G3 vs. G4	0.501578	0.564832

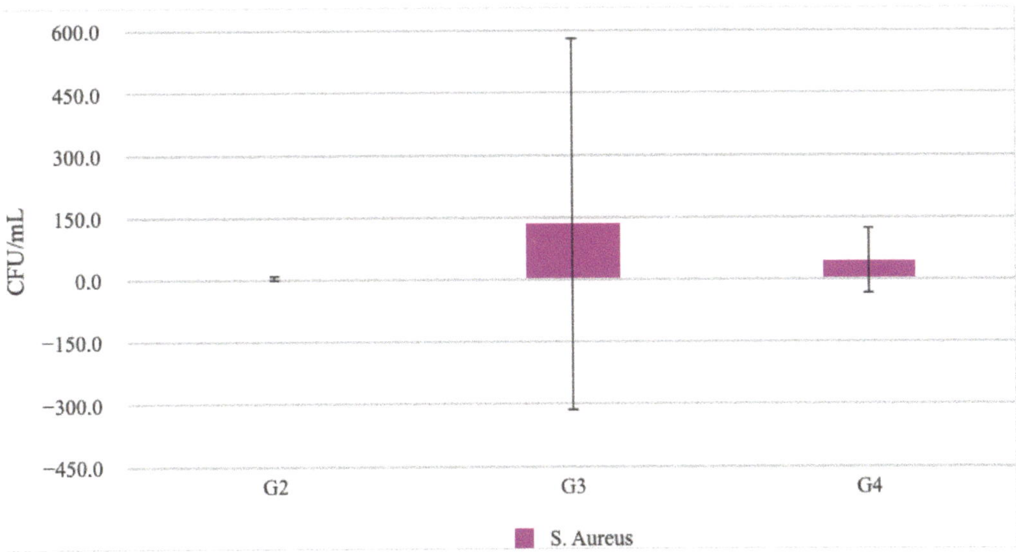

Figure 7. Comparison of the success of individual protocols in the eradication of *S. aureus*: G2—positive control, G3—hydrogen peroxide, and G4—laser-activated hydrogen peroxide.

Descriptive statistics of CFU/mL for all protocols regarding *C. albicans* are shown in Table 3. In the negative control, the highest proportion of colonies per mL was present at 4.06×10^5, followed by the group disinfected with hydrogen peroxide (G3) with 24 CFU/mL, the positive control (G2) with 20 CFU/mL, and the laser-activated hydrogen peroxide group (G4) with the smallest proportion of 5 CFU/mL.

Table 3. Descriptive statistics of CFU/mL of *C. albicans* under different protocols.

CFU/mL *C. albicans*	**Sample Size**	**Medium Value**	**Minimal Value**	**Maximal Value**	**Standard Deviation**
Negative control	20	406,240	0.00	4,000,000	959,808
Positive control	20	20	0.00	200	61.56
3% H_2O_2	20	24	0.00	200	60.73
Laser + H_2O_2	20	5	0.00	40	11.00

CFU/mL—colony-forming units per milliliter.

Furthermore, there was no statistically significant difference in *C. albicans* eradication between the individually tested protocols: positive control (G2), hydrogen peroxide (G3), and laser-activated hydrogen peroxide (G4) (Table 4 and Figure 8).

Table 4. A performance comparison of different protocols in the reduction of the *C. albicans* CFU.

Multiple Comparisons p-Values (2-Tailed); CFU/mL *C. albicans* Independent (Grouping) Variable: Protocol Kruskal–Wallis Test: H (3, N = 80) = 36.32679 p = 0.0000				
Protocol	**Negative Control R:63.30**	**Positive Control R:30.10**	**3% H_2O_2 R:36.00**	**Laser + H_2O_2 R:32.60**
Negative control		0.000037	0.001219	0.000177
Positive control	0.000037		1.000000	1.000000
3% H_2O_2	0.001219	1.000000		1.000000
Laser + H_2O_2	0.000177	1.000000	1.000000	

CFU/mL—colony-forming units per milliliter.

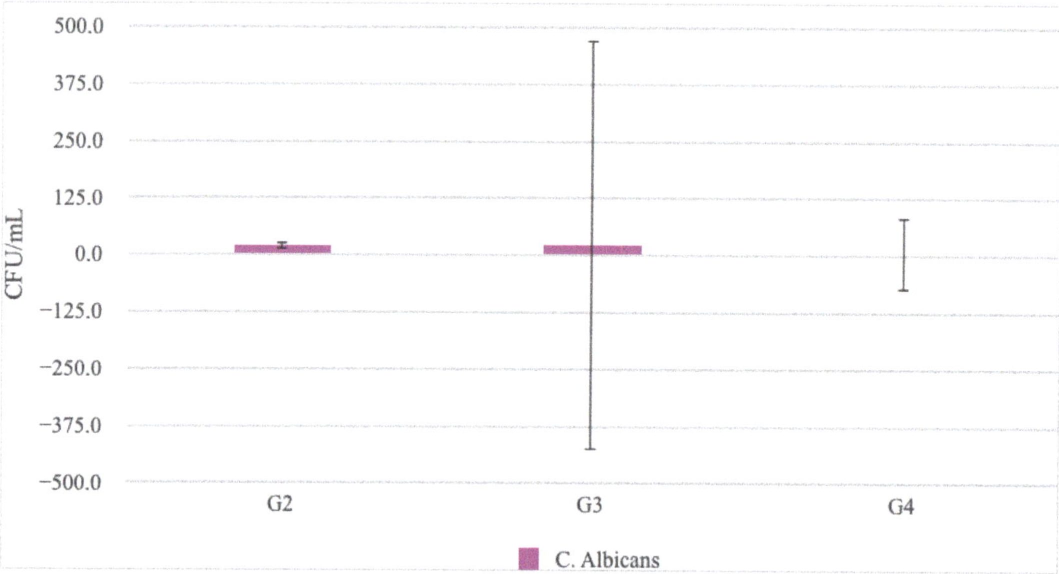

Figure 8. Comparison of the success of individual protocols in the eradication of *C. albicans*: G2—positive control, G3—3% hydrogen peroxide, and G4—laser-activated 3% hydrogen peroxide.

4. Discussion

This research primarily examined the possibility of improving the antimicrobial efficiency of 3% H_2O_2 in peri-implantitis therapy, using diode laser activation with a wavelength of 445 nm, compared to non-activated H_2O_2 and the current gold standard of 0.2% CHX [23]. All tested groups showed a massive, significant reduction in microbial CFU count when compared to the negative control. As for the hypothesis, the first hypothesis was only partially accepted. Although the laser-activated group clearly showed better results in the microbial CFU count, there was no statistically significant difference after the statistical analysis, likely due to the sample size. The second hypothesis was accepted, since the analysis revealed comparable results between CHX and H_2O_2 groups. H_2O_2 is one of the disinfectants used in dental medicine. Oxygen radicals (ROS) generated by the application of H_2O_2 have a high affinity for bacteria and the extracellular matrix of bacteria, and a strong oxidizing effect. Additionally, they ensure the creation of active oxygen foam that also has mechanical cleaning properties. ROS (hydroxyl radicals) can theoretically damage healthy tissue cells, but ROS generated by the application of 3% H_2O_2 have a rapid breakdown into water and oxygen molecules [24]. Also, the incidence of ROS, which occurs with photoactivation of 3% H_2O_2, was examined, and it was shown that the formation of ROS stops when the photoactivation with the laser device stops [25]. In the research of Wiedmer et al. [26], a comparison of the effects of CHX and H_2O_2 on the biofilm mass on the implant surface was shown. It has been proven that H_2O_2 has a strong effect on actively removing the mass of biofilm, while CHX has a slightly weaker effect on reducing the said mass, but it certainly has a bactericidal effect on the microbial community and prevents the regrowth of bacteria. The current study showed a roughly equal effectiveness of CHX and H_2O_2 in eradicating *S. aureus* and *C. albicans*. This study was conducted on the surface of real dental implants rather than titanium discs, in order to simulate in vivo conditions and obtain more realistic results [27,28]. *C. albicans* and *S. aureus* cultures were chosen for research due to the simple cultivation, control of the microorganisms themselves, and also due to the fact that they create polymicrobial biofilms ("early colonizers") and actively participate in peri-implant infections [13,29].

Diode lasers belong to the newer group of techniques combating the causative agent of periodontitis [30]. They demonstrate a disinfection effectiveness in two main ways: through photothermal and/or photodynamic therapy. The effectiveness of photothermal therapy is based on the thermal energy generated by the laser during radiation emission. The generated thermal energy transfers indirectly through the environment or directly on the bacterial cell, acting fatally on it [31]. Regarding the energy levels of lasers, proper energy dosage is crucial: too little energy will not be sufficient to eradicate enough microorganisms and too much energy could cause damage to the neighboring tissues [32]. Usual antimicrobial energy levels in peri-implant decontamination studies range from 0.5 to 3 W in continuous (CW) or pulsed-mode (energy emission types) [32–37]. Photodynamic therapy with diode lasers is based on the light activation of the photosensitive agent in presence of oxygen with a low-energy (sub-ablative) laser beam. There is no heating, and thus no risk of damage to the surrounding tissues [38]. The principles of photodynamic therapy are: photosensitive molecules (the dye) bind to target microorganisms on the implant surface, which is then irradiated with a light of a certain wavelength in the presence of oxygen. Light-excited photosensitizers undergo type I (electron transfer) and/or type II (energy transfer) reactions to produce reactive oxygen species (ROS), resulting in the disruption of the bacterial cell wall and/or normal metabolism, leading to bacterial cell damage or death [38,39]. The described mechanisms do not harm human cells, as these cells have mechanisms to survive oxidative stress (catalase and superoxide dismutase enzymes).

The development of microbial resistance to PDT is not probable, as the bactericidal effect is achieved through the action of oxygen radicals on the cellular components of the micro-organisms [38].

In the current research, a new diode laser, with a wavelength of 445 nm, was used to test the possibility of activating 3% H_2O_2. H_2O_2 exhibits maximum absorption at wavelengths of approximately 400 nm. The idea was to try to activate H_2O_2 photodynamically, and also photothermally (by heating) at the same time; therefore, the energy settings were set to 1 W CW for 60 s, with constant movement of the laser optical fiber (1 mm/s) along the surface of the implant, from a distance of approximately 2 mm. The thermal effect of these settings was not measured, although the settings, combined with the constant displacement/freehand movement of the laser fiber tip in a wet environment (movement prevents excessive heat accumulation on one place and allows cooling microbreaks), could be considered as relatively safe and were therefore chosen for research [32–36]. Nevertheless, additional testing should be done regarding the temperature rise in the specific protocol and its effects on the surrounding tissue.

Chemical reactions in the activated H_2O_2 environment were not assessed. Potential activation/antimicrobial augmentation was assessed only via the CFU count, which is a more basic, simpler approach. Complex chemical analysis should be done in future similar studies to closer analyze the interaction of this specific laser wavelength and H_2O_2.

The specified wavelength of the diode laser was first examined in the concrete disinfection protocol by Katalinic et al. [40], however not in peri-implant conditions, but on endodontic intracanal biofilms composed of E. faecalis, *C. albicans*, and *S. aureus*, where promising results were obtained. The same laser was then tested in peri-implantitis therapy, but as part of photodynamic therapy with 0.1% riboflavin, where a positive antimicrobial effect was also demonstrated [41]. By reviewing the literature available to the authors, it is not possible to find research similar to the current one. However, research examining the impact of laser energy on H_2O_2 and various pathogens exists. In the study by Ikai et al. [42], the authors analyzed the effect of activated hydrogen peroxide on cultures of S. mutans, A. actinomycetemcomitans, E. faecalis, and *S. aureus*. It has been proven that hydrogen peroxide, activated by a laser with a wavelength of 405 nm, has the ability to eliminate all four pathogens within 3 min, which is not the case when using lasers or H_2O_2 as independent treatments. Photoactivation of H_2O_2, with the aim of improving the antimicrobial effect, was also examined in the work of Mahdi et al. in 2015 [21]. The authors proved a stronger disinfection potential when H_2O_2 was activated by LED light with a wavelength of 440–480 nm, compared to non-activated H_2O_2. In contrast to the current work, the photoactivation was performed with LED light and not with a diode laser that has monochromatic, coherent, and collimated light radiation. In two scientific papers, Odor et al. [43,44] investigated the antimicrobial effect of hydroxyl radicals produced by diode laser photoactivation, in combination with conventional mechanical periodontitis therapy. A diode laser with a wavelength of 940 nm and a power of 1 W was used. The laser-activated H_2O_2 group showed the best results. In relation to the other research mentioned, that research was not done in vitro, but in vivo, on patients with periodontitis, and the effect on bacterial cultures was examined differently from that in the current research.

From all presented studies, it is possible to conclude that laser activation of H_2O_2 has a strong, positive and potentially clinically relevant antimicrobial effect, but the results cannot be directly compared with the current study due to too many differences in the design of the study (microorganisms tested, energy settings of the laser, different wavelengths, different presentation of the obtained data, etc.). Additional research is needed to determine the exact impact of 445 nm laser energy on H_2O_2 and most efficient laser energy settings, leading to a safe and useful clinical decontamination protocol.

5. Conclusions

The conducted research provides a preliminary insight into the protocol for treating the surface of dental implants with a combination of agents that have not been described in the literature so far. Statistical analysis revealed a significant difference between all three disinfection protocols compared to the negative control. However, in the mutual comparison of the results of the three disinfection protocols, there were no statistically significant results, although the laser-activated H_2O_2 group showed better antimicrobial results compared to non-activated 3% H_2O_2. Within the inherent limitations of this study, it is possible to conclude that all three disinfection protocols are equally powerful in the treatment of S.aureus and C.albicans biofilms. For further evaluation of efficiency, new research is needed on a larger number of implants, testing more energy settings, thermal effects of the settings and, certainly, in-clinical, "in vivo" set-ups.

Author Contributions: Formal analysis, I.K., I.S., L.M. and A.B.; writing—original draft preparation, I.K., writing—review and editing, D.G. and M.V. All authors have read and agreed to the published version of the manuscript.

Funding: This research received no external funding.

Institutional Review Board Statement: Not applicable.

Informed Consent Statement: Not applicable.

Data Availability Statement: The data presented in this study are available on request from the corresponding author.

Conflicts of Interest: The authors declare no conflict of interest.

References

1. Tarnow, D.P. Commentary: Replacing Missing Teeth With Dental Implants: A Century of Progress. *J. Periodontol.* **2014**, *85*, 1475–1477. [CrossRef] [PubMed]
2. Buser, D.; Sennerby, L.; De Bruyn, H. Modern implant dentistry based on osseointegration: 50 years of progress, current trends and open questions. *Periodontology 2000* **2017**, *73*, 7–21. [CrossRef]
3. Albrektsson, T.; Donos, N. Implant survival and complications. The Third EAO consensus conference 2012. *Clin. Oral Implant. Res.* **2012**, *23*, 63–65. [CrossRef]
4. Srinivasan, M.; Vazquez, L.; Rieder, P.; Moraguez, O.; Bernard, J.-P.; Belser, U.C. Survival rates of short (6 mm) micro-rough surface implants: A review of literature and meta-analysis. *Clin. Oral Implant. Res.* **2014**, *25*, 539–545. [CrossRef]
5. Smeets, R.; Henningsen, A.; Jung, O.; Heiland, M.; Hammächer, C.; Stein, J.M. Definition, etiology, prevention and treatment of pe-ri-implantitis—A review. *Head Face Med.* **2014**, *10*, 34. [CrossRef]
6. Atieh, M.A.; Alsabeeha, N.H.; Faggion, C.M., Jr.; Duncan, W.J. The frequency of peri-implant diseases: A systematic review and me-ta-analysis. *J Periodontol.* **2013**, *84*, 1586–1598. [CrossRef]
7. Rams, T.E.; Degener, J.E.; Van Winkelhoff, A.J. Antibiotic resistance in human peri-implantitis microbiota. *Clin. Oral Implant. Res.* **2014**, *25*, 82–90. [CrossRef]
8. Salvi, G.E.; Fürst, M.M.; Lang, N.P.; Persson, G.R. One-year bacterial colonization patterns of Staphylococcus aureus and other bacteria at implants and adjacent teeth. *Clin. Oral Implant. Res.* **2008**, *19*, 242–248. [CrossRef]
9. Negrini, T.C.; Koo, H.; Arthur, R.A. Candida-Bacterial Biofilms and Host-Microbe Interactions in Oral Diseases. *Adv. Exp. Med. Biol.* **2019**, *1197*, 119–141.
10. Otto, M. Staphylococcal biofilms. In *Gram-Positive Pathogens*, 3rd ed.; Fischetti, V.A., Novick, R.P., Eds.; John Wiley & Sons, Ltd.: Hoboken, NJ, USA, 2019; pp. 699–711.
11. Harriott, M.M.; Noverr, M.C. Importance of Candida–bacterial polymicrobial biofilms in disease. *Trends Microbiol.* **2011**, *19*, 557–563. [CrossRef]
12. De-La-Torre, J.; Quindós, G.; Marcos-Arias, C.; Marichalar-Mendia, X.; Gainza, M.L.; Eraso, E.; Acha-Sagredo, A.; Aguirre-Urizar, J.M. Oral Candida colonization in patients with chronic periodontitis. Is there any relationship? *Rev. Iberoam. Micol.* **2018**, *35*, 134–139. [CrossRef]
13. Rokaya, D.; Srimaneepong, V.; Wisitrasameewon, W.; Humagain, M.; Thunyakitpisal, P. Peri-implantitis Update: Risk Indicators, Diagnosis, and Treatment. *Eur. J. Dent.* **2020**, *14*, 672–682. [CrossRef]
14. Scarano, A.; Nardi, G.; Murmura, G.; Rapani, M.; Mortellaro, C. Evaluation of the Removal Bacteria on Failed Titanium Implants After Irradiation With Erbium-Doped Yttrium Aluminium Garnet Laser. *J. Craniofacial Surg.* **2016**, *27*, 1202–1204. [CrossRef]
15. Memè, L.; Sartini, D.; Pozzi, V.; Emanuelli, M.; Strappa, E.M.; Bittarello, P.; Bambini, F.; Gallusi, G. Epithelial Biological Response to Machined Titanium vs. PVD Zirconium-Coated Titanium: An In Vitro Study. *Materials* **2022**, *15*, 7250. [CrossRef]
16. Alovisi, M.; Carossa, M.; Mandras, N.; Roana, J.; Costalonga, M.; Cavallo, L.; Pira, E.; Putzu, M.G.; Bosio, D.; Roato, I.; et al. Disinfection and Biocompatibility of Titanium Surfaces Treated with Glycine Powder Airflow and Triple Antibiotic Mixture: An In Vitro Study. *Materials* **2022**, *15*, 4850. [CrossRef]
17. Al-Hashedi, A.A.; Laurenti, M.; Benhamou, V.; Tamimi, F. Decontamination of titanium implants using physical methods. *Clin. Oral Implant. Res.* **2017**, *28*, 1013–1021. [CrossRef]
18. Louropoulou, A.; Slot, D.E.; Van Der Weijden, F.A. Titanium surface alterations following the use of different mechanical instruments: A systematic review. *Clin. Oral Implant. Res.* **2012**, *23*, 643–658. [CrossRef]
19. De Araújo Nobre, M.; Capelas, C.; Alves, A.; Almeida, T.; Carvalho, R.; Antunes, E.; Oliveira, D.; Cardador, A.; Maló, P. Non-surgical treatment of peri-implant pathology. *Int. J. Dent. Hyg.* **2006**, *4*, 84–90. [CrossRef]
20. Marshall, M.V.; Cancro, L.P.; Fischman, S.L. Hydrogen Peroxide: A Review of Its Use in Dentistry. *J. Periodontol.* **1995**, *66*, 786–796. [CrossRef]
21. Mahdi, Z.; Habiboallh, G.; Mahbobeh, N.N.; Mina, Z.J.; Majid, Z.; Nooshin, A. Lethal effect of blue light-activated hydrogen peroxide, curcumin and erythrosine as potential oral photosensitizers on the viability of Porphyromonas gingivalis and Fusobacterium nucleatum. *Laser Ther.* **2015**, *24*, 103–111. [CrossRef]
22. Ehrenfest, D.M.D.; Del Corso, M.; Kang, B.S.; Leclercq, P. Identification card and codification of the chemical and morphological characteristics of 62 dental implant surfaces. Part 3: Sand-blasted/acid-etched (SLA type) and related surfaces (Group 2A, main subtractive process). *Poseido* **2014**, *2*, 37–55.
23. Heitz-Mayfield, L.J.; Mombelli, A. The Therapy of Peri-implantitis: A Systematic Review. *Int. J. Oral Maxillofac. Implant.* **2014**, *29*, 325–345. [CrossRef] [PubMed]
24. Chapple, I.L.C.; Matthews, J.B. The role of reactive oxygen and antioxidant species in periodontal tissue destruction. *Periodontology 2000* **2007**, *43*, 160–232. [CrossRef] [PubMed]

25. Jaimes, E.A.; Sweeney, C.; Raij, L. Effects of the Reactive Oxygen Species Hydrogen Peroxide and Hypochlorite on Endothelial Nitric Oxide Production. *Hypertension* **2001**, *38*, 877–883. [CrossRef]
26. Wiedmer, D.; Petersen, F.C.; Lönn-Stensrud, J.; Tiainen, H. Antibacterial effect of hydrogen peroxide-titanium dioxide suspensions in the decontamination of rough titanium surfaces. *Biofouling* **2017**, *33*, 451–459. [CrossRef]
27. Teughels, W.; Van Assche, N.; Sliepen, I.; Quirynen, M. Effect of material characteristics and/or surface topography on biofilm development. *Clin. Oral Implants Res.* **2006**, *17*, 68–81. [CrossRef] [PubMed]
28. Gustumhaugen, E.; Lönn-Stensrud, J.; Scheie, A.A.; Lyngstadaas, S.P.; Ekfeldt, A.; Taxt-Lamolle, S. Effect of chemical and mechanical debridement techniques on bacterial re-growth on rough titanium surfaces: An in vitro study. *Clin. Oral Implants Res.* **2014**, *25*, 707–713. [CrossRef]
29. Charalampakis, G.; Belibasakis, G.N. Microbiome of peri-implant infections: Lessons from conventional, molecular and metagenomic analyses. *Virulence* **2015**, *6*, 183–187. [CrossRef] [PubMed]
30. Marotti, J.; Tortamano, P.; Cai, S.; Ribeiro, M.S.; Franco, J.E.M.; De Campos, T.T. Decontamination of dental implant surfaces by means of photodynamic therapy. *Lasers Med. Sci.* **2013**, *28*, 303–309. [CrossRef]
31. Olivi, G.; De Moor, R.; DiVito, E. *Lasers in Endodontics: Scientific Background and Clinical Application*, 1st ed.; Springer International Publishing: Cham, Switzerland; Heidelberg, Germany; New York, NY, USA; Dordrecht, The Netherlands; London, UK, 2016.
32. Deppe, H.; Ahrens, M.; Behr, A.V.; Marr, C.; Sculean, A.; Mela, P.; Ritschl, L.M. Thermal effect of a 445 nm diode laser on five dental implant systems: An in vitro study. *Sci. Rep.* **2021**, *11*, 20174. [CrossRef] [PubMed]
33. Malmqvist, S.; Liljeborg, A.; Qadri, T.; Johannsen, G.; Johannsen, A. Using 445 nm and 970 nm Lasers on Dental Implants—An In Vitro Study on Change in Temperature and Surface Alterations. *Materials* **2019**, *12*, 3934. [CrossRef] [PubMed]
34. Valente, N.A.; Calascibetta, A.; Patianna, G.; Mang, T.; Hatton, M.; Andreana, S. Thermodynamic Effects of 3 Different Diode Lasers on an Implant-Bone Interface: An Ex-Vivo Study With Review of the Literature. *J. Oral Implantol.* **2017**, *43*, 94–99. [CrossRef] [PubMed]
35. Matys, J.; Flieger, R.; Dominiak, M. Effect of diode lasers with wavelength of 445 and 980 nm on a temperature rise when un-covering implants for second stage surgery: An ex-vivo study in pigs. *Adv. Clin. Exp. Med.* **2017**, *26*, 687–693. [CrossRef]
36. Geminiani, A.; Caton, J.G.; Romanos, G.E. Temperature change during non-contact diode laser irradiation of implant surfaces. *Lasers Med. Sci.* **2012**, *27*, 339–342. [CrossRef]
37. Romanos, G.E. Diode laser soft-tissue surgery: Advancements aimed at consistent cutting, improved clinical outcomes. *Compend. Contin. Educ. Dent.* **2013**, *34*, 752–758. [PubMed]
38. Cieplik, F.; Deng, D.; Crielaard, W.; Buchalla, W.; Hellwig, E.; Al-Ahmad, A.; Maisch, T. Antimicrobial photodynamic therapy. What we know and what we dont. *Crit. Rev. Microbiol.* **2018**, *44*, 571–589. [CrossRef] [PubMed]
39. Kwiatkowski, S.; Knap, B.; Przystupski, D.; Saczko, J.; Kędzierska, E.; Knap-Czop, K.; Kotlińska, J.; Michel, O.; Kotowski, K.; Kulbacka, J. Photodynamic therapy—Mechanisms, photosensitizers and combinations. *Biomed Pharmacother.* **2018**, *106*, 1098–1107. [CrossRef]
40. Katalinić, I.; Budimir, A.; Bošnjak, Z.; Jakovljević, S.; Anić, I. The photo-activated and photo-thermal effect of the 445/970 nm diode laser on the mixed biofilm inside root canals of human teeth in vitro: A pilot study. *Photodiagnosis Photodyn. Ther.* **2019**, *26*, 277–283. [CrossRef]
41. Morelato, L.; Budimir, A.; Smojver, I.; Katalinić, I.; Vuletić, M.; Ajanović, M.; Gabrić, D. A Novel Technique for Disinfection Treatment of Contaminated Dental Implant Surface Using 0.1% Riboflavin and 445 nm Diode Laser-An In Vitro Study. *Bioengineering* **2022**, *9*, 308. [CrossRef]
42. Ikai, H.; Nakamura, K.; Shirato, M.; Kanno, T.; Iwasawa, A.; Sasaki, K.; Niwano, Y.; Kohno, M. Photolysis of Hydrogen Peroxide, an Effective Disinfection System via Hydroxyl Radical Formation. *Antimicrob. Agents Chemother.* **2010**, *54*, 5086–5091. [CrossRef]
43. Odor, A.A.; Bechir, E.S.; Forna, D.A. Effect of Hydrogen Peroxide Photoactivated Decontamination Using 940 nm Diode Laser in Periodontal Treatment: A Pilot Study. *Photobiomodulation Photomed. Laser Surg.* **2020**, *38*, 614–624. [CrossRef] [PubMed]
44. Odor, A.A.; Bechir, E.S.; Violant, D.; Badea, V. Antimicrobial Effect of 940 nm Diode Laser based on Photolysis of Hydrogen Peroxide in the Treatment of Periodontal Disease. *Rev. Chim.* **2018**, *69*, 2081–2088. [CrossRef]

biomedicines

MDPI

Systematic Review

Photodynamic Therapy Can Modulate the Nasopharyngeal Carcinoma Microenvironment Infected with the Epstein–Barr Virus: A Systematic Review and Meta-Analysis

Diógenes Germano Fornel [1], Túlio Morandin Ferrisse [2], Analú Barros de Oliveira [3] and Carla Raquel Fontana [1,*]

1 Department of Clinical Analysis, School of Pharmaceutical Sciences, UNESP-São Paulo State University, Araraquara 14801-902, SP, Brazil
2 Department of Dental Materials and Prosthodontics, School of Dentistry, UNESP-São Paulo State University, Araraquara 14801-903, SP, Brazil
3 Department of Orthodontics and Pediatric Dentistry, School of Dentistry, UNESP-São Paulo State University, Araraquara 14801-903, SP, Brazil
* Correspondence: carla.fontana@unesp.br

Abstract: Nasopharyngeal carcinoma is a malignancy from epithelial cells predominantly associated with the Epstein–Barr virus (EBV) infection, and it is responsible for 140,000 deaths annually. There is a current need to develop new strategies to increase the efficacy of antineoplastic treatment and reduce side effects. Thus, the present study aimed to perform a systematic review and meta-analysis of the ability of photodynamic therapy (PDT) to modulate the tumor microenvironment and PDT efficacy in nasopharyngeal carcinoma treatment. The reviewers conducted all steps in the systematic review. PubMed, Science Direct, Scopus, Scielo, Lilacs, EMBASE, and the Cochrane library databases were searched. The OHAT was used to assess the risk of bias. Meta-analysis was performed with a random-effects model ($\alpha = 0.05$). Nasopharyngeal carcinoma cells treated with PDT showed that IL-8, IL-1α, IL-1β, LC3BI, LC3BII, MMP2, and MMP9 levels were significantly higher than in groups that did not receive PDT. NF-κB, miR BART 1-5p, BART 16, and BART 17-5p levels were significantly lower in the PDT group than in the control group. Apoptosis levels and the viability of nasopharyngeal carcinoma cells (>70%) infected with EBV were effective after PDT. This treatment also increased LMP1 levels (0.28–0.50/$p < 0.05$) compared to the control group. PDT showed promising results for efficacy in killing nasopharyngeal carcinoma cells infected with EBV and modulating the tumor microenvironment. Further preclinical studies should be performed to validate these results.

Keywords: photodynamic therapy; nasopharyngeal carcinoma; Epstein–Barr virus; systematic review; meta-analysis

1. Introduction

Head and neck cancer comprises a heterogeneous group of malignancies with distinguished etiological factors [1]. The human microbiome has received attention as a risk factor for head and neck cancer [2]. In this regard, *Streptococcus anginous*, *Streptococcus mitis*, *Streptococcus oral*, *Streptococcus gordonii*, *Capnocytophaga gingivalis*, *Prevotella melaninogenica* and *Porphyromonas gingivalis* are examples of microorganisms related to the oncogenesis of head and neck cancer [2]. Moreover, the human papillomavirus (HPV) and Epstein–Barr virus (EBV) strongly correlate to oro- and nasopharyngeal carcinomas, respectively [2].

The EBV (human herpes virus type 4) belongs to the *Gammaherpesvirinae* family and is responsible for infecting more than 90% of the world's population [3]. Although EBV causes infectious mononucleosis, it is associated with carcinomas and lymphoma carcinogenesis, resulting in 1% of global cancers. Approximately 140,000 people die each year from EBV-related tumors [4].

Nasopharyngeal carcinoma is a malignancy from epithelial cells that extends across the nasopharynx surface, and EBV is one of its etiological factors [4–6]. EBV infection divides the level of differentiation of nasopharyngeal epithelial cells into type II (non-keratinizing) and type III (undifferentiated). Hence, a non-keratinizing tumor is predominantly associated with EBV infection [7]. Nasopharyngeal carcinoma prevails in men, with the highest incidence in North Africa and Southeast Asia, particularly in southern China and eastern Malaysia [4]. Additionally, EBV viral load; history of chronic diseases (ear, nose, or throat); genetic factors; environmental exposure; and the consumption of alcohol, tobacco, salted fish, dairy, and lipids are among the etiological factors of nasal carcinoma [4,8].

The latent membrane protein type 1 (LMP1) is present during the latent phase of EBV and is considered an oncogenic protein. Moreover, LMP1 has a higher number of polymorphisms than other genes [9], and it can induce the production of tumors, possibly due to the functional similarity of the anti-protein with tumor necrosis factor (TNF-α), CD40, and tumor necrosis factor type I (TNF-1) [10]. LMP1 can also up-regulate anti-apoptotic genes, down-regulate metastasis suppressors, and promote angiogenesis, pro-inflammatory cytokine activation, and epithelial cell morphology changes [11].

The conventional treatment for nasopharyngeal carcinoma is radiotherapy, chemotherapy, and surgical resection [8]. However, the overall survival rate of patients affected by this carcinoma is still low, and the bad prognoses have remained independent of treatment [12]. Furthermore, other chemotherapy drugs, such as docetaxel, cisplatin, and fluorouracil associated with chemoradiotherapy, have not improved the five-year overall survival and progression-free survival rates [13]. That highlights the absence of new effective therapeutic options for treating these patients [12]. Additionally, conventional treatment (chemotherapy and radiotherapy) is associated with different adverse events, such as skin hyperpigmentation, fatigue, nausea, leukopenia, anemia, hepatotoxicity, and diarrhea [14].

Photodynamic therapy (PDT) involves applying a photosensitizer, followed by a light source in a specific wavelength, to a target tissue [6,15]. After the photosensitizer is sensitized by irradiation, reactive oxygen species are produced, causing cytotoxicity and indirect destruction of tumor cells due to vascular damage [6]. Therefore, PDT might be a promising approach for treating nasopharyngeal carcinoma. Regarding the high prevalence of EBV infection associated with the development of nasopharyngeal carcinoma, the present study aimed to perform a systematic review and meta-analysis of the ability of PDT to modulate the tumor microenvironment and PDT effectiveness in killing nasopharyngeal carcinoma cells infected with EBV.

2. Materials and Methods

2.1. Protocol and Registration

The present systematic review and meta-analysis were performed according to the Preferred Reporting Items for Systematic Reviews (PRISMA) statement [16]. The study was registered in the Open Science Framework (OSF) (registration DOI: 10.17605/OSF.IO/ACUVG).

2.2. Data Extraction and Study Question

The research question was based on the PICO strategy for systematic exploratory reviews [16], where P = nasopharyngeal carcinoma cells infected with EBV; I = photodynamic therapy (PDT); C = PDT associated with another therapy, the absence or application of another treatment instead of PDT, or nasopharyngeal carcinoma cells not infected with EBV; O = the primary outcome was chemokine and interleukin levels and the second one was the viability of nasopharyngeal carcinoma cells infected with EBV and LMP1 levels. The present study aimed to answer the following focused questions: What is the efficacy of PDT in reducing nasopharyngeal carcinoma cells infected with EBV? Moreover, can PDT modulate the inflammatory microenvironment in this tumor infected with EBV?

2.3. Eligibility Criteria

The inclusion criteria for the systematic review were in vitro studies that used PDT to treat nasopharyngeal carcinoma cells infected with EBV and cell lines from humans. There was no restriction on types of language, photosensitizer, and cell line. The exclusion criteria were observational studies and clinical trials in humans; book chapters; letters to the editor; conference abstracts; theses; dissertations; case reports; and studies with nasopharyngeal carcinoma cells not infected with EBV, without evaluating the photoinactivation of EBV, and with cell lines from animals.

2.4. Search Strategy

Two independent examiners were calibrated in a previous pilot study to perform the steps for article selection. The electronic search was performed in PubMed, Science Direct, Scopus, Scielo, Lilacs, EMBASE, and the Cochrane library databases. The search words were (((Epstein-Barr) OR (Epstein-Barr virus)) OR (EBV)) AND (Photodynamic therapy). The Kappa calibration ($0.87/p < 0.01$) between the examiners was an "almost perfect" agreement. Mendeley Reference Software was used to detect and eliminate duplicates. After the eligibility step, the data were extracted from the selected articles, analyzed, and discussed. Any disagreement during the process was solved before proceeding to the next steps by reaching a consensus. The following data were extracted from the included studies: first name of the author, year of publication, study design, cell lineage, sample size, evaluated group, photosensitizer, wavelength (nanometers), irradiation time (minutes), incubation time of the photosensitizer, light dose, and main results.

2.5. Risk of Bias Assessment

This step used the OHAT Rob Rating tool adapted for in vitro studies [17,18]. There were four answer alternatives for each question: (i) definitely low (++) there is direct evidence to affirm the answer to the question; (ii) (+) there is indirect evidence to affirm the answer to the question; (iii) (−) there is indirect evidence to respond negatively to the question; (iv) (−−) there is direct evidence to respond negatively to the question. The question "Were there no other potential threats to internal validity?" referred to a bias related to statistical approaches (sample size calculation, normality and homoscedasticity evaluations, and inferential text details) [19].

2.6. Meta-Analysis

The meta-analysis used the random-effects model, the standard mean difference in effect measurement. A forest plot was made to evaluate the results better. The trim-and-fill method was used to detect publication and meta-analysis biases. Heterogenicity levels above 50% were considered high ($I^2 > 50\%$). R software, version 3.6.3, and Rstudio with the "META" package were used to conduct quantitative approaches ($\alpha = 0.05$) and build the graphs.

3. Results

3.1. Search Results

The flowchart in Figure 1 summarizes the article selection process. The electronic search yielded 203 articles. Accordingly, 175 articles remained for selection. After title and abstract screenings, 168 articles were excluded because they did not meet the eligibility criteria. Seven studies were eligible for a full-text evaluation. After the full-text assessment, the same seven articles were included in the qualitative analysis, and three were included in the meta-analysis. Four studies were excluded from the quantitative analysis because they did not report the sample size or precisely report the outcome measurement (Supplementary Materials).

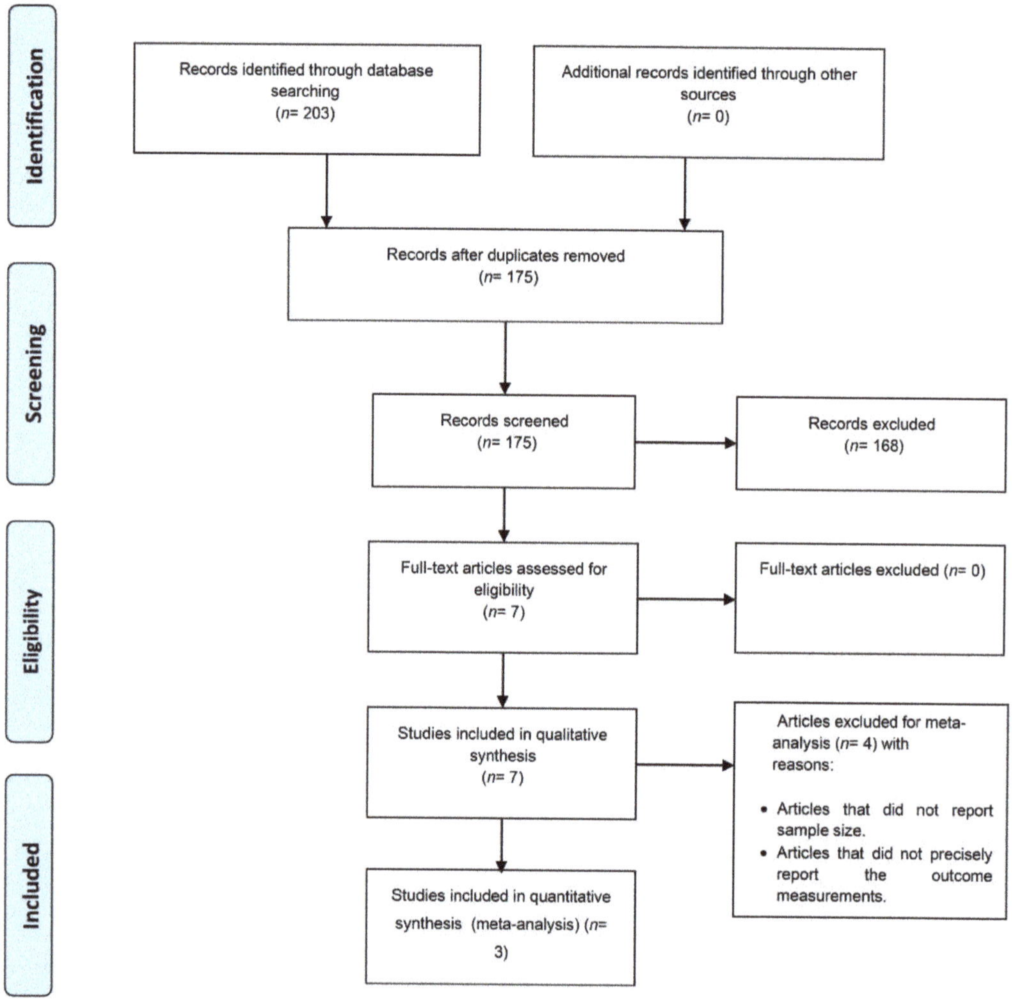

Figure 1. Flowchart summarizing the search results of the present study.

3.2. *Synthesis of Results*

The articles included in the present systematic review and meta-analysis ranged in publication dates from 2002 to 2020 [20–26] (Table 1). The cell lines most frequently used were CNE-2 (57.15%) and C666-1 (57.15%), followed by HK-1 (42.86%). Numerous photosensitizers were used, but two studies evaluated the same one (FosPeg®®) [23,25]. The wavelength varied from 585 to 685 J/cm^2 and the light dose from 0.25 to 20 J/cm^2. The photosensitizer incubation time ranged from three to 24 h, and four hours was the most frequently used [20,23,26]. Only one study reported the irradiation time [24].

Table 1. Data extracted from the articles included in the present systematic review and meta-analysis.

#	Author	Study Design	Cell Line	Sample Size	Evaluated Group	Photosensitizer	Wavelength (nm)	Irradiation Time (minutes)	Photosensitizer Incubation Time	Light Dose	Results
1	Du et al., 2002 [20]	In vitro	HK-1 CNE-2	6	G1: PDT G2: no PDT	Hypericin	585	ND	4 h (HK-1) 6 h (CNE-2)	0.5 J/cm^2	**IL-8 (pg/mL)** HK-1/ G1: 168.80 ± 7.93 HK-1/G2: 130.80 ± 5.80 CNE-2/ G1:71.15 ± 9.81 CNE-2/G2: 60.09 ± 2.01
2	Koon et al., 2010 [21]	In vitro	HK-1	ND	G1: HK-1 (EBV+) G2: HK-1 (EBV-) G3: control (no PDT + EBV+)	Zn-BC-AM	682	ND	24 h	0.25–1.0 J/cm^2	**Apoptosis (PI)** HK-1 (EBV+): 80% HK-1 (EBV−): 60% **IL-1α (pg/mL)** HK-1 (EBV+): 6300 ± 250 HK-1 (EBV−): 3301 ± 500 Control/G3: 1350 ± 250 **IL-1β (pg/mL)** HK-1 (EBV+): 92 ± 5 HK-1 (EBV−): 55 ± 5 Control/G3: 18 ± 2 **IL-8 (pg/mL)** HK-1 (EBV+): 15 ± 1 HK-1 (EBV−): 0 ± 0 Control/G3: 430 ± 25
3	Li et al., 2010 [22]	In vitro	c666-1 CNE-2	3	G1: c666-1 (EBV+) G2: CNE-2 (EBV-)	HMME (7(12)-(1-methoxyethyl)-12(7)-(1-hydroxyethyl)-3,8,13,17-tetramethyl-21H,23H-porphin-2,18-dipropionic)	630	ND	3 h	0.6–14.4 J/cm^2	**Phototoxicity (clonogenic assay)** There were significant and similar results for G1 and G2, particularly when the intracellular uptake of HMME was balanced between the groups.
4	Wu et al., 2013 [23]	In vitro	c666-1 HK-1 CNE-2	3	G1: c666-1 (EBV+) G2: HK-1 (EBV-) G3: CNE-2 (EBV-)	FosPeg	630	ND	4 h	3.0 J/cm^2	**Cytotoxicity (MTT)** c666-1: 69% HK-1: 77% CNE-2: 84% **LMP1 mRNA expression** c666-1: 8 ± 1.5 (PDT+) c666-1: 1 ± 0.0 (PDT−) **EBV-miR-BART 1-5p** c666-1: 0.75 ± 0.1 (PDT+) c666-1: 1.0 ± 0.0 (PDT−) **EBV-miR-BART 16** c666-1: 0.6 ± 0.25 (PDT+) c666-1: 1.0 ± 0.0 (PDT−) **EBV-miR-BART 17-5p** c666-1: 0.75 ± 0.1 (PDT+) c666-1: 1.0 ± 0.1 (PDT−) **LMP1 protein expression** c666-1: 1.35 ± 0.15 (PDT+) c666-1: 1.0 ± 0.1 (PDT−)

Table 1. *Cont.*

#	Author	Study Design	Cell Line	Sample Size	Evaluated Group	Photosensitizer	Wavelength (nm)	Irradiation Time (minutes)	Photosensitizer Incubation Time	Light Dose	Results
5	Peng et al., 2017 [24]	In vitro	NPC 5-8F NPC 6-10B	ND	G1: PDT G2: PDT + Lovastatin	Photosan II	630	1	24 h	10 J/cm^2	**Cell viability (Alamar blue)** There were significant results for Lovastatin + PDT for both cell lines.
6	Wu et al., 2020 (a) [25]	In vitro	c666-1	3	G1: 2D culture G2: 3D culture (MCL and MCS)	FosPeg	652	ND	24 h	20 J/cm^2	**Cell viability (MTT)** 2D: 95 ± 5% MCL: 60 ± 10% MCS: 70% **Apoptosis (Annexin V)** 2D: 30.6 ± 7.7 MCL: 31.0 ± 7.4 MCS: 27.6 ± 7.0 **Necrosis (Annexin V)** 2D: 16.3 ± 8.6 MCL: 9.8 ± 10.6 MCS: 13.5 ± 3.2 **LC3BI protein expression** 2D: 1.5 ± 1.0 MCL: 1.4 ± 1.2 MCS: 0.8 ± 0.5 **LC3BII protein expression** 2D: 1.8 ± 1.0 MCL: 1.25 ± 0.8 MCS: 0.8 ± 0.5 **LMP1 protein expression** 2D: 0.9 ± 0.25 MCL: 1.25 ± 1.0 MCS: 2.0 ± 1.25 **MMP2 protein expression** 2D: 0.7 ± 0.15 MCL: 1.2 ± 0.25 MCS: 1.5 ± 1.0 **MMP9 protein expression** 2D: 0.7 ± 0.15 MCL: 2.2 ± 0.75 MCS: 1.5 ± 0.65 **ABCB1 protein expression** 2D: 0.5 ± 0.25 MCL: 1.5 ± 0.65 MCS: 1 ± 0.8 **ABCC1 protein expression** 2D: 1.0 ± 0.25 MCL: 2.3 ± 1.2 MCS: 1.8 ± 0.1 **ABCG2 protein expression** 2D: 1.7 ± 0.5 MCL: 1.5 ± 0.5 MCS: 1.8 ± 2.0

Table 1. *Cont.*

#	Author	Study Design	Cell Line	Sample Size	Evaluated Group	Photosensitizer	Wavelength (nm)	Irradiation Time (minutes)	Photosensitizer Incubation Time	Light Dose	Results
7	Wu et al., 2020 (b) [26]	In vitro	c666-1 CNE-2	3	G1: c666-1 (EBV+) G2: CNE-2 (EBV-)	H-ALA (5-aminolevulinic acid hexyl ester)	630	ND	4 h	2–4 J/cm^2	**Cytotoxicity (MTT)** G1: 70% G2: 80% **LMP1 protein expression** G1: 1.5 ± 0.0 Control: 1.0 ± 0 **EGRF protein expression** G1: 0.75 ± 0.16 G2: 0.6 ± 0.0 Control: 1.0 ± 0.0 **p-EGRF protein expression** G1: 0.5 ± 0.3 G2: 0.8 ± 0.16 Control: 1.0 ± 0.0 **NF-κB protein expression** G1: 0.8 ± 0.25 G2: 0.8 ± 0.16 Control: 1.0 ± 0.0

EBV: Epstein-Barr virus; NPC: nasopharyngeal carcinoma; PDT: photodynamic therapy. Evaluated groups: G1—group 1, G2—group 2, G3—group 3; PDT: hypericin-based photodynamic therapy; Photosensitizer: hematoporphyrin monomethyl ether. Irradiation time: minutes; ND: not documented; Light dose: J—joules; Results: H-ALA—5-aminolevulinic acid hexyl derivative, MCL—liquid overlay method with agarose base, MCS—hanging drop method, PpIX—protoporphyrin IX. After PDT in nasopharyngeal carcinoma cells infected with EBV, the IL-8 [20], IL-1α, IL-1β [21], LMP1 [23,25,26], LC3BI, LC3BII, MMP2, and MMP9 [25] levels were not higher than control groups. ABCB1, ABCC1, and ABCG2 [25] levels did not show significant results compared to the control group. There were significant results for apoptosis levels [21] and the viability of nasopharyngeal carcinoma cells infected with EBV. However, NF-κB protein expression decreased after PDT for the same group compared to nasopharyngeal carcinoma cells not infected with EBV [26]. miR BART 1-5p, BART 16, and BART 17-5p levels also decreased [23].

3.3. Risk of Bias Assessment

The primary source of bias in all included articles referred to blinding (were research personnel blind to the study group during the investigation?/was the outcome assessment reliable, including the blinding of evaluators?) and details about statistical approaches (were there no other potential threats to internal validity?) (Table 2).

Table 2. Risk of bias analysis according to the OHAT Rob Rating tool adapted to assess the risk of bias of in vitro studies included in the systematic review.

Questions/Studies	Du et al., 2002 [20]	Koon et al., 2010 [21]	Li et al., 2010 [22]	Wu et al., 2013 [23]	Peng et al., 2017 [24]	Wu et al., 2020 (a) [25]	Wu et al., 2020 (b) [26]
Was the administered dose or exposure level adequately randomized?	++	++	++	++	++	++	++
Were study group allocations adequately concealed?	++	++	++	++	++	++	++
Were the experimental conditions identical across study groups?	++	++	++	++	++	++	++
Were research personnel blind to the study group during the investigation?	–	–	–	–	–	–	–
Were outcome data complete without attrition or exclusion from the analysis?	++	++	++	++	++	++	++
Was the exposure characterization reliable?	++	++	++	++	++	++	++
Was the outcome assessment reliable (including the blinding of evaluators)?	–	–	–	–	–	–	–
Were there no other potential threats to internal validity?	– –	– –	– –	– –	– –	– –	– –

++ = direct evidence to affirm the question; – = indirect evidence to respond negatively to the question; – – = direct evidence to respond negatively to the question.

3.4. Meta-Analysis

The meta-analysis was only possible for LMP1 levels [23,25,26]. Thus, the experimental group included nasopharyngeal carcinoma cells infected with EBV treated with PDT, and the control group consisted of nasopharyngeal carcinoma cells not infected with EBV and without receiving PDT. PDT increased LMP1 levels (mean difference (MD) = 0.28/95% confidence interval (CI) = 0.01–0.56/I^2 = 90%) (Figure 2a). After detecting the publication bias with the trim-and-fill method and correlating the meta-analysis, MD was 0.50 [0.28–0.72], but the heterogenicity level remained high (I^2 = 90%) (Figure 2b).

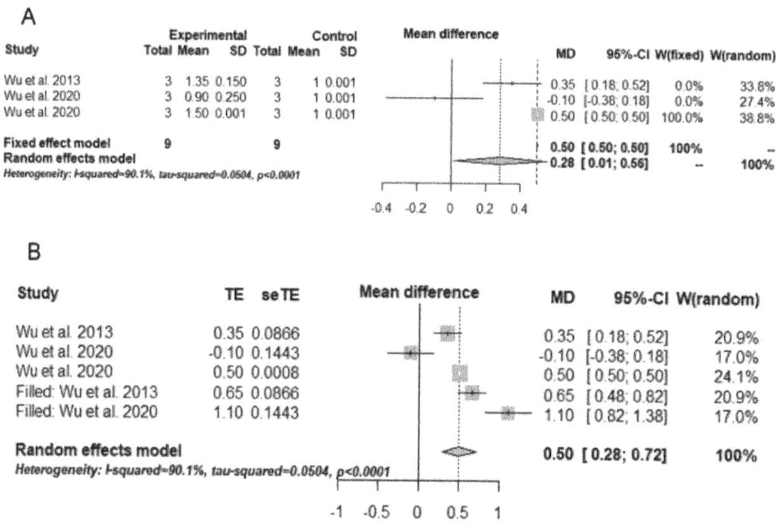

Figure 2. Results of the quantitative approaches used in the present study. (**A**) Meta-analysis illustrated in a forest plot for LMP1 levels with significant results. (**B**) Trim-and-fill results illustrated

in a forest plot. This statistical test detected publication and meta-analysis biases, which were corrected. The experimental group included nasopharyngeal carcinoma cells infected with EBV and treated with PDT, and the control group consisted of nasopharyngeal carcinoma cells not infected with EBV and without receiving PDT. MD = mean difference; SD = standard deviation; CI = confidence interval; TE = estimated mean; seTE = estimated standard deviation [23,25,26].

4. Discussion

According to the World Health Organization, there are three pathological subtypes of nasopharyngeal carcinomas: keratinized squamous, non-keratinized, and basaloid squamous [27]. Nonetheless, the non-keratinized subtype represents more than 95% of cases in endemic areas and is predominantly associated with EBV infection [7,27]. This tumor is related to a remarkable geographical distribution. Thus, there are other risk factors for developing nasopharyngeal carcinoma in addition to EBV infection, such as host genetics and environmental aspects (e.g., salted fish consumption) [28,29].

Tumor-derived epithelial cells are susceptible to ionizing radiation, which explains why radiotherapy is the primary treatment modality for non-metastatic nasopharyngeal carcinoma [28]. Chemotherapy combined with radiotherapy is essential and highly indicated for advanced locoregional diseases. Patients with metastatic nasopharyngeal carcinoma are a heterogeneous group, and although chemotherapy is the mainstay treatment modality at this stage, individualized treatment is increasingly required [28,30]. A high dose of radiation or chemotherapy causes acute and later side effects. Oral mucositis, dermatitis, xerostomia, and dysphagia are the main acute toxicities associated with radiotherapy, and xerostomia, sensorineural hearing loss, osteoradionecrosis, trismus, and hormonal dysfunction (e.g., hypothyroidism) are the described later effects [28]. Hematological discrepancies are the main toxicities when administering chemotherapy in nasopharyngeal carcinoma patients [28].

Despite advances in radiotherapy and chemotherapy for treating nasopharyngeal carcinoma, the overall survival rate is still poor and the side effects reduce the quality of life of patients diagnosed with this tumor [12,28,31]. In this scenario, PDT is a treatment option for different cancers [19,32–34]. There is a current lack of clinical trials evaluating the effectiveness of PDT for nasopharyngeal carcinoma, which is the main reason for developing this present systematic review on in vitro studies.

Nasopharyngeal carcinoma cells infected with EBV and treated with PDT significantly increased IL-8, IL-1α, and IL-1β levels, resulting in cell death. PDT also increased IL-1α and IL-1β levels in nasopharyngeal carcinoma cells without EBV infection, but at a lower rate than tumors infected with EBV [21]. Moreover, applying PDT only to the tumor did not affect IL-8 levels. In return, apoptosis [21] and cytotoxicity levels from PDT [23,26] were similar in nasopharyngeal carcinoma cells regardless of EBV infection. That is promising because the oxidative damage of PDT can cause different cell deaths and immunological response pathways depending on EBV infection in nasopharyngeal carcinoma cells.

In particular, IL-8 is an inflammatory mediator mainly related to necroptosis [35] that can play a different role in cancer. However, IL-8 can recruit innate immune cells, starting an immunological response against cancer [36]. That is a strength of PDT over other traditional cancer treatments because PDT can initiate immunogenic cell death accompanied by the exposure and release of damage-associated molecular patterns (DAMPs) [37]. In the context of the tumor microenvironment, cancer cells can die by apoptosis, necroptosis, and autophagy, along with inflammatory molecule release that may modulate the immunological response against cancer [37]. Furthermore, the role of IL-8 in PDT differs from radiotherapy, in which IL-8 induces an epithelial–mesenchymal transition [38] and tumor cell repopulation after radiotherapy via RIP1/RIP3/MLKL/JNK/IL-8 pathways [35], leading to a poor prognosis for cancer patients.

IL-1 is a pro-inflammatory cytokine that participates in nasopharyngeal carcinoma development and is recognized as an oncogenic factor for this tumor. High IL-1 levels are normal in nasopharyngeal carcinoma and are stimulated by T cells infiltrated in the tumor and lipopolysaccharides (LPS) [39]. In this scenario, LPS-containing Gram-negative bacteria can stimulate resident macrophages via TLR4 for TNF and IL-1 secretion, inducing cell proliferation and tumorigenesis [40].

Despite the role of IL-1 in nasopharyngeal processes, there is a lack of knowledge about the action of this cytokine in the tumor microenvironment after treatment. The caspase-1/NLRP3/IL-1 pathway regards inflammasome formation and pyroptosis, which stimulate the activation of inflammatory processes and modulation of immune responses [41]. Thus, PDT might induce nasopharyngeal cell death via inflammasome formation, but further studies should be designed to evaluate this point better.

EBV-induced carcinogenesis in nasopharyngeal carcinoma can explain the differences in inflammatory responses of nasopharyngeal carcinoma cells infected with EBV or not after PDT. There are three EBV latent phases, distinguished by viral antigen expression. Type I shows EBNA1 (EBV nuclear antigen 1) expression, type II presents EBNA1 and LMP1, LMP2, and EBERs (EBV-encoded small RNA), and type III includes a high production of EBNA1,2, LMP1, LMP2, and EBRs [42,43].

LMP1 is an oncogenic protein that stimulates the expression of the epidermal growth factor receptor (EGFR), promoting cell growth by activating the MAP kinase pathway [44–46]. In epithelial cells, LMP1 inhibits P53-mediated induction of apoptosis and induces lymphocyte sensibilization to TGF-beta, tempering the immune response against cancer cells [47]. In other words, the distinguished carcinogenesis pathways related to different etiological factors (EBV positive and EBV negative) may be considered the reason for different responses obtained after PDT for nasopharyngeal carcinoma.

LMP1 can also activate oncogenic signaling pathways, causing tumor invasion, metastasis, anti-apoptosis ability, and inhibition of squamous cell differentiation [28]. Higher LMP1 levels indicate a poor prognosis for nasopharyngeal carcinoma patients. The meta-analysis showed higher LMP1 levels in nasopharyngeal carcinoma cells treated with PDT than those not treated with PDT. However, as LMP1 function depends on the activation of NF-κB and STAT3 pathways [48] and the group treated with PDT showed lower NF-κB levels, PDT might make LMP1 dysfunctional.

In other words, higher LMP1 levels can represent a direct response to cell death by oxidative damage from PDT; therefore, LMP1 would not run an oncogenic pathway because NF-κB levels decreased. Although our study did not evaluate this finding, we might also hypothesize that, as LMP1 function depends on the NF-κB and STAT3 pathways, PDT might modulate LMP1 function via the STAT3 pathways in persistent cancer cells. However, future studies should assess these very pathways better. It is also essential to balance the benefits between increased LMP1 levels via the NF-κB and STAT3 pathways and PDT efficiency in killing nasopharyngeal carcinoma cells (>70%). The impact of higher LMP1 levels on clinical trials remains to be analyzed.

BART-miRNAs are transcription factors that potentiate tumor growth, cooperate in immune attack escape, and strengthen anti-apoptosis ability [28]. PDT reduced BART-miRNAs levels, indicating that PDT could also positively modulate tumor-mediated factors, possibly improving the prognosis of nasopharyngeal carcinoma patients.

MMP2 and MMP9 are a family of proteolytic enzymes implicated in the invasion and metastasis of numerous cancers because they degrade extracellular matrix components [49]. MMP2 and MMP9 overexpression is associated with higher tumor grades. Concomitantly, MMP2 overexpression is associated with a higher risk of cancer metastasis, and MMP9 overexpression correlates to lymph node metastasis [49,50]. However, MMP2, MMP9, and other proteolytic enzymes in local inflammations from PDT can cooperate in tissue damage by facilitating a reduction in tumor volume [51].

The role of autophagy in cancer development, growth, invasion, and metastasis has recently been highlighted. In this context, microtubule-associated protein 1B light chain 3B

(LC3B) is one of the most studied proteins, and its overexpression is associated with a poor prognosis [52]. PDT triggers autophagy in tumor cells by suppressing AKT-mTOR signaling or up-regulating the AMPK pathway [53,54]. Thus, LC3B overexpression indicates that tumor cells underwent autophagy. Related concerns should be mentioned because surviving tumor cells can obtain resistance over PDT by inhibiting autophagy pathways [55].

The effect of PDT on cancer cells is related to apoptosis or necrosis, and autophagy is an intracellular degradation pathway that can participate in pro-survival or pro-death mechanisms. Thus, autophagy often monitors cellular death by PDT as an attempt to survive oxidative damage [55]. Furthermore, autophagy inhibition can decrease anti-apoptotic proteins, promoting survival and tumor adaption against PDT [56]. The precise mechanisms that can unbalance autophagy from running toward pro-death cells are pivotal for improving the clinical outcomes of cancer patients treated with PDT [41]. Thus, autophagy after PDT with different photosensitizers should be further investigated.

The ATP-binding cassette (ABC) transporters are transmembrane proteins that utilize ATP to transport/efflux diverse compounds across cellular membranes [57]. Among these proteins, ABCB1, ABCC1, and ABCG2 can transport numerous chemotherapy drugs outside cells, causing chemoresistance [58]. PDT could not affect ATP protein levels in nasopharyngeal carcinoma cells infected with EBV, which can be a good response because of the absence of tumor resistance by photodamage in this pathway. The present study adapted the OHAT Rob Rating tool to assess the risk of bias for in vitro studies [17,18]. Most included articles showed a higher risk of bias related to blinding. Although blinding is not frequently used for in vitro studies and is highly required in randomized clinical trials, this methodological approach was accepted, considering that effect size estimates may be overrated. Blinding can also eliminate the observation bias [59].

The findings of the present systematic review and meta-analysis should be understood with caution because only seven articles were included in the systematic review and three in the meta-analysis. Moreover, three of the seven articles were published by the same research group, representing a limitation for the present study. Hence, the studies were conducted with different cell lines, such as CNE-2, C666-1, and HK-1. CNE-2 is a poorly differentiated nasopharyngeal carcinoma epithelioid cell line from a primary tumor biopsy in China [60]. The C666-1 cell line represents an undifferentiated nasopharyngeal carcinoma carrying EBV in long-term cultures [61]. The HK-1 originated from a recurrent (after radiotherapy) differentiated nasopharyngeal carcinoma [62]. These differences can represent distinguished molecular signatures that could cause differences in tumor microenvironment responses after PDT and impact PDT efficacy in killing tumor cells.

In summary, PDT can modulate the tumor microenvironment of nasopharyngeal carcinoma cells and is an efficient treatment against these cells when infected with EBV. However, these findings should be investigated in animals and previous preclinical studies.

5. Conclusions

PDT is a promising approach as a treatment for nasopharyngeal carcinoma cells infected with EBV because it can modulate the tumor microenvironment. It also showed significant results in killing nasopharyngeal carcinoma cells infected with EBV. Nevertheless, PDT can easily be associated with other treatments for this tumor.

Supplementary Materials: The following supporting information can be downloaded at: https://www.mdpi.com/article/10.3390/biomedicines11051344/s1. The reference [63] is cited in the Supplementary File.

Author Contributions: Conceptualization, T.M.F., A.B.d.O., and C.R.F.; methodology, T.M.F. and A.B.d.O.; software, T.M.F.; validation, D.G.F., T.M.F., A.B.d.O., and C.R.F.; formal analysis, D.G.F., T.M.F., and A.B.d.O.; investigation, D.G.F., T.M.F., A.B.d.O., and C.R.F.; data curation: T.M.F. and C.R.F.; writing—original draft preparation, D.G.F. and T.M.F.; writing—review and editing, T.M.F., A.B.d.O., and C.R.F.; visualization, D.G.F., T.M.F., A.B.d.O., and C.R.F.; supervision, T.M.F., A.B.d.O., and C.R.F.; project administration, T.M.F., A.B.d.O., and C.R.F., funding acquisition, T.M.F., A.B.d.O., and C.R.F. All authors have read and agreed to the published version of the manuscript.

Funding: This research was supported by the São Paulo Research Foundation (FAPESP #2013/07276-1) and scholarships (FAPESP grants #2022/12828-2, #2021/01191-0, #2020/07110-0). FAPESP was not involved in the study design, data collection, analysis, and interpretation.

Institutional Review Board Statement: Not applicable.

Informed Consent Statement: Not applicable.

Data Availability Statement: Not applicable.

Conflicts of Interest: The authors declare no conflict of interest.

References

1. Torre, L.A.; Bray, F.; Siegel, R.L.; Ferlay, J.; Lortet-Tieulent, J.; Jemal, A. A global cancer statistic, 2012. *CA Cancer J. Clin.* **2015**, *65*, 87–108. [CrossRef] [PubMed]
2. Orlandi, E.; Iacovelli, N.A.; Tombolini, V.; Rancati, T.; Polimeni, A.; De Cecco, L.; Valdagni, R.; De Felice, F. Potential role of microbiome in oncogenesis, outcome prediction and therapeutic targeting for head and neck cancer. *Oral Oncol.* **2019**, *99*, 104453. [CrossRef]
3. Chang, C.M.; Yu, K.J.; Mbulaiteye, S.M.; Hildesheim, A.; Bhatia, K. The extent of genetic diversity of Epstein-Barr virus and its geographic and disease patterns: A need for reappraisal. *Virus Res.* **2009**, *143*, 209–221. [CrossRef]
4. Bakkalci, D.; Jia, Y.; Winter, J.R.; Lewis, J.E.; Taylor, G.S.; Stagg, H.R. Risk factors for Epstein Barr virus-associated cancers: A systematic review, critical appraisal, and mapping of the epidemiological evidence. *J. Glob. Health* **2020**, *10*, 010405. [CrossRef]
5. Sham, J.; Choy, D.; Wei, W.; Ng, M.H.; Zong, Y.-S.; Lin, Z.-X.; Guo, Y.-Q.; Luo, Y. Detection of subclinical riasopharyngeal carcinoma by fibreoptic endoscopy and multiple biopsy. *Lancet* **1990**, *335*, 371–374. [CrossRef]
6. Renaud, S.; Lefebvre, A.; Mordon, S.; Moralès, O.; Delhem, N. Novel Therapies Boosting T Cell Immunity in Epstein Barr Virus-Associated Nasopharyngeal Carcinoma. *Int. J. Mol. Sci.* **2020**, *21*, 4292. [CrossRef] [PubMed]
7. Young, L.S.; Dawson, C.W. Epstein-Barr virus and nasopharyngeal carcinoma. *Chin. J. Cancer* **2014**, *33*, 581–590. [CrossRef] [PubMed]
8. Hutajulu, S.H.; Kurnianda, J.; Tan, B.I.; Middeldorp, J.M. Therapeutic implications of Epstein-Barr virus infection for the treatment of nasopharyngeal carcinoma. *Ther. Clin. Risk Manag.* **2014**, *10*, 721–736. [CrossRef]
9. Banko, A.; Miljanovic, D.; Lazarevic, I.; Cirkovic, A. A Systematic Review of Epstein–Barr Virus Latent Membrane Protein 1 (LMP1) Gene Variants in Nasopharyngeal Carcinoma. *Pathogens* **2021**, *10*, 1057. [CrossRef] [PubMed]
10. Edilova, M.I.; Abdul-Sater, A.A.; Watts, T.H. TRAF1 Signaling in Human Health and Disease. *Front. Immunol.* **2018**, *9*, 2969. [CrossRef]
11. da Costa, V.G.; Marques-Silva, A.C.; Moreli, M.L. The Epstein-Barr virus latent membrane protein-1 (LMP1) 30-bp deletion and XhoI-polymorphism in nasopharyngeal carcinoma: A meta-analysis of observational studies. *Syst. Rev.* **2015**, *4*, 46. [CrossRef]
12. Ahn, M.-J.; Chirovsky, D.; Kuyas, H.; Auclair, V.; Abounit, S.; Joo, S.; Shah, R.; Yang, M.-H. Global longitudinal assessment of treatment outcomes in recurrent/metastatic nasopharyngeal carcinoma: GLANCE-NPC study. *Futur. Oncol.* **2021**, *17*, 2015–2025. [CrossRef]
13. Ou, D.; Blanchard, P.; El Khoury, C.; De Felice, F.; Even, C.; Levy, A.; Nguyen, F.; Janot, F.; Gorphe, P.; Deutsch, E.; et al. Induction chemotherapy with docetaxel, cisplatin and fluorouracil followed by concurrent chemoradiotherapy or chemoradiotherapy alone in locally advanced non-endemic nasopharyngeal carcinoma. *Oral Oncol.* **2016**, *62*, 114–121. [CrossRef]
14. Zong, J.; Liu, Y.; Liang, Q.; Xu, H.; Chen, B.; Guo, Q.; Xu, Y.; Hu, C.; Pan, J.; Lin, S. Administration of oral maintenance chemotherapy for 1 year following definitive chemoradiotherapy may improve the survival of patients with stage N3 nasopharyngeal carcinoma. *Oral Oncol.* **2021**, *118*, 105313. [CrossRef] [PubMed]
15. Stoker, S.D.; van Diessen, J.N.A.; de Boer, J.P.; Karakullukcu, B.; Leemans, C.R.; Tan, I.B. Current Treatment Options for Local Residual Nasopharyngeal Carcinoma. *Curr. Treat. Options Oncol.* **2013**, *14*, 475–491. [CrossRef]
16. Shamseer, L.; Moher, D.; Clarke, M.; Ghersi, D.; Liberati, A.; Petticrew, M.; Shekelle, P.; Stewart, L.A.; PRISMA-P Group. Preferred reporting items for systematic review and meta-analysis protocols (PRISMA-P) 2015: Elaboration and explanation. *BMJ* **2015**, *349*, g7647. [CrossRef]
17. NTP-OHAT. *OHAT Risk of Bias Rating Tool for Human and Animal Studies*; Office of Health Assessment and Translation: Rockville, MD, USA, 2015.
18. NTP-OHAT. *Handbook for Conducting a Literature-Based Health Assessment Using OHAT Approach for Systematic Review and Evidence Integration*; National Toxicology Program–Office of Health Assessment and Translation: Rockville, MD, USA, 2019.
19. Ferrisse, T.M.; de Oliveira, A.B.; Surur, A.K.; Buzo, H.S.; Brighenti, F.L.; Fontana, C.R. Photodynamic therapy associated with nanomedicine strategies for treatment of human squamous cell carcinoma: A systematic review and meta-analysis. *Nanomedicine* **2022**, *40*, 102505. [CrossRef]

20. Du, H.; Bay, B.H.; Mahendran, R.; Olivo, M. Endogenous expression of interleukin-8 and interleukin-10 in nasopharyngeal carcinoma cells and the effect of photodynamic therapy. *Int. J. Mol. Med.* **2002**, *10*, 73–76. [CrossRef]
21. Koon, H.K.; Lo, K.W.; Leung, K.N.; Lung, M.L.; Chang, C.C.; Wong, R.N.; Leung, W.N.; Mak, N.K. Photodynamic therapy-mediated modulation of inflammatory cytokine production by Epstein-Barr virus-infected nasopharyngeal carcinoma cells. *Cell Mol. Immunol.* **2010**, *7*, 323–326. [CrossRef] [PubMed]
22. Li, B.; Chen, Z.; Liu, L.; Huang, Z.; Huang, Z.; Xie, S. Differences in sensitivity to HMME-mediated photodynamic therapy between EBV+ C666-1 and EBV- CNE2 cells. *Photodiagnosis Photodyn. Ther.* **2010**, *7*, 204–209. [CrossRef] [PubMed]
23. Wu, R.W.; Chu, E.S.; Huang, Z.; Xu, C.S.; Ip, C.W.; Yow, C.M. FosPeg® PDT alters the EBV miRNAs and LMP1 protein expression in EBV positive nasopharyngeal carcinoma cells. *J. Photochem. Photobiol. B* **2013**, *127*, 114–122. [CrossRef]
24. Peng, Y.; He, G.; Tang, D.; Xiong, L.; Wen, Y.; Miao, X.; Hong, Z.; Yao, H.; Chen, C.; Yan, S.; et al. Lovastatin Inhibits Cancer Stem Cells and Sensitizes to Chemo- and Photodynamic Therapy in Nasopharyngeal Carcinoma. *J. Cancer* **2017**, *8*, 1655–1664. [CrossRef]
25. Wu, R.W.K.; Chu, E.S.M.; Yuen, J.W.M.; Huang, Z. Comparative study of FosPeg® photodynamic effect on nasopharyngeal carcinoma cells in 2D and 3D models. *J. Photochem. Photobiol. B* **2020**, *210*, 111987. [CrossRef]
26. Wu, R.W.K.; Chu, E.S.M.; Yow, C.M.N. Evaluation of the effect of 5-aminolevulinic acid hexyl ester (H-ALA) PDT on EBV LMP1 protein expression in human nasopharyngeal cells. *Photodiagnosis Photodyn. Ther.* **2020**, *30*, 101801. [CrossRef]
27. Wang, H.-Y.; Chang, Y.-L.; To, K.-F.; Mai, H.-Q.; Feng, Y.-F.; Chang, E.T.; Wang, C.-P.; Kam, M.K.M.; Cheah, S.-L.; Lee, M.; et al. A new prognostic histopathologic classification of nasopharyngeal carcinoma. *Chin. J. Cancer* **2016**, *35*, 41. [CrossRef]
28. Chen, Y.P.; Chan, A.T.C.; Le, Q.T.; Blanchard, P.; Sun, Y.; Ma, J. Nasopharyngeal carcinoma. *Lancet* **2019**, *394*, 64–80. [CrossRef]
29. Tsao, S.W.; Yip, Y.L.; Tsang, C.M.; Pang, P.S.; Lau, V.M.; Zhang, G.; Lo, K.W. Etiological factors of nasopharyngeal carcinoma. *Oral Oncol.* **2014**, *50*, 330–338. [CrossRef]
30. Bossi, P.; Chan, A.; Licitra, L.; Trama, A.; Orlandi, E.; Hui, E.; Halámková, J.; Mattheis, S.; Baujat, B.; Hardillo, J.; et al. Nasopharyngeal carcinoma: ESMO-EURACAN Clinical Practice Guidelines for diagnosis, treatment, and follow-up†. *Ann. Oncol.* **2021**, *32*, 452–465. [CrossRef] [PubMed]
31. Mayor, S. Side-effects of cancer drugs are under-reported in trials. *Lancet Oncol.* **2015**, *16*, e107. [CrossRef] [PubMed]
32. Fontana, L.C.; Pinto, J.G.; Vitorio, G.D.S.; Ferreira, I.; Pacheco-Soares, C.; Mamone, L.A.; Strixino, J.F. Photodynamic effect of protoporphyrin IX in gliosarcoma 9l/lacZ cell line. *Photodiagnosis Photodyn. Ther.* **2022**, *37*, 102669. [CrossRef]
33. Mkhobongo, B.; Chandran, R.; Abrahamse, H. The Role of Melanoma Cell-Derived Exosomes (MTEX) and Photodynamic Therapy (PDT) within a Tumor Microenvironment. *Int. J. Mol. Sci.* **2021**, *22*, 9726. [CrossRef]
34. Vallecorsa, P.; Di Venosa, G.; Gola, G.; Sáenz, D.; Mamone, L.; MacRobert, A.J.; Ramírez, J.; Casas, A. Photodynamic therapy of cutaneous T-cell lymphoma cell lines mediated by 5-aminolevulinic acid and derivatives. *J. Photochem. Photobiol. B* **2021**, *221*, 112244. [CrossRef] [PubMed]
35. Wang, Y.; Zhao, M.; He, S.; Luo, Y.; Zhao, Y.; Cheng, J.; Gong, Y.; Xie, J.; Wang, Y.; Hu, B.; et al. Necroptosis regulates tumor repopulation after radiotherapy via RIP1/RIP3/MLKL/JNK/IL8 pathway. *J. Exp. Clin. Cancer Res.* **2019**, *38*, 461. [CrossRef] [PubMed]
36. Beltrán Hernández, I.; Yu, Y.; Ossendorp, F.; Korbelik, M.; Oliveira, S. Preclinical and Clinical Evidence of Immune Responses Triggered in Oncologic Photodynamic Therapy: Clinical Recommendations. *J. Clin. Med.* **2020**, *9*, 333. [CrossRef] [PubMed]
37. Huis In 't Veld, R.V.; Heuts, J.; Ma, S.; Cruz, L.J.; Ossendorp, F.A.; Jager, M.J. Current Challenges and Opportunities of Photodynamic Therapy against Cancer. *Pharmaceutics* **2023**, *15*, 330. [CrossRef]
38. Teijeira, A.; Garasa, S.; Ochoa, M.C.; Villalba, M.; Olivera, I.; Cirella, A.; Eguren-Santamaria, I.; Berraondo, P.; Schalper, K.A.; de Andrea, C.E.; et al. IL8, Neutrophils, and NETs in a Collusion against Cancer Immunity and Immunotherapy. *Clin. Cancer Res.* **2021**, *27*, 2383–2393. [CrossRef]
39. Allen, D.Z.; Aljabban, J.; Silverman, D.; McDermott, S.; Wanner, R.A.; Rohr, M.; Hadley, D.; Panahiazar, M. Meta-Analysis illustrates possible role of lipopolysaccharide (LPS)-induced tissue injury in nasopharyngeal carcinoma (NPC) pathogenesis. *PLoS ONE* **2021**, *16*, e0258187. [CrossRef]
40. Yang, Y.; Liao, Q.; Wei, F.; Li, X.; Zhang, W.; Fan, S.; Shi, L.; Li, X.; Gong, Z.; Ma, J.; et al. LPLUNC1 inhibits nasopharyngeal carcinoma cell growth via down-regulation of the MAP kinase and cyclin D1/E2F pathways. *PLoS ONE* **2013**, *8*, e62869. [CrossRef]
41. Di Paolo, N.C.; Shayakhmetov, D.M. Interleukin 1α and the inflammatory process. *Nat. Immunol.* **2016**, *17*, 906–913. [CrossRef] [PubMed]
42. Perri, F.; Della Vittoria Scarpati, G.; Giuliano, M.; D'Aniello, C.; Gnoni, A.; Cavaliere, C.; Licchetta, A.; Pisconti, S. Epstein-Barr virus infection and nasopharyngeal carcinoma: The other side of the coin. *Anticancer Drugs* **2015**, *26*, 1017–1025. [CrossRef]
43. Perri, F.; Sabbatino, F.; Ottaiano, A.; Fusco, R.; Caraglia, M.; Cascella, M.; Longo, F.; Rega, R.A.; Salzano, G.; Pontone, M.; et al. Impact of Epstein Barr Virus Infection on Treatment Opportunities in Patients with Nasopharyngeal Cancer. *Cancers* **2023**, *15*, 1626. [CrossRef] [PubMed]
44. Dawson, C.W.; Rickinson, A.B.; Young, L.S. Epstein-Barr virus latent membrane protein inhibits human epithelial cell differentiation. *Nature* **1990**, *344*, 777–780. [CrossRef] [PubMed]
45. Fhraeus, R.; Rymo, L.; Rhim, J.S.; Klein, G. Morphological transformation of human keratinocytes expressing the LMP gene of Epstein-Barr virus. *Nature* **1990**, *345*, 447–449. [CrossRef] [PubMed]

46. Miller, W.E.; Earp, H.S.; Raab-Traub, N. The Epstein-Barr virus latent membrane protein 1 induces expression of the epidermal growth factor receptor. *J. Virol.* **1995**, *69*, 4390–4398. [CrossRef]
47. Kieff, E.; Rickinson, A.B. Epstein-Barr virus and its replication. In *Field's Virology*; Knipe, D.M., Howley, P.M., Eds.; Lippincott/Williams & Wilkins: Philadelphia, PA, USA, 2001; Volume 2, pp. 2511–2573.
48. Lo, A.K.; Dawson, C.W.; Lung, H.L.; Wong, K.L.; Young, L.S. The Role of EBV-Encoded LMP1 in the NPC Tumor Microenvironment: From Function to Therapy. *Front. Oncol.* **2021**, *11*, 640207. [CrossRef]
49. Pietruszewska, W.; Bojanowska-Poźniak, K.; Kobos, J. Matrix metalloproteinases MMP1, MMP2, MMP9 and their tissue inhibitors TIMP1, TIMP2, TIMP3 in head and neck cancer: An immunohistochemical study. *Otolaryngol. Pol.* **2016**, *70*, 32–43. [CrossRef]
50. Jiang, H.; Li, H. Prognostic values of tumoral MMP2 and MMP9 overexpression in breast cancer: A systematic review and meta-analysis. *BMC Cancer* **2021**, *21*, 149. [CrossRef]
51. Chen, Y.; Ma, H.; Wang, W.; Zhang, M. A size-tunable nanoplatform: Enhanced MMP2-activated chemo-photodynamic immunotherapy based on biodegradable mesoporous silica nanoparticles. *Biomater. Sci.* **2021**, *9*, 917–929. [CrossRef]
52. Giatromanolaki, A.; Koukourakis, M.I.; Georgiou, I.; Kouroupi, M.; Sivridis, E. LC3A, LC3B and Beclin-1 Expression in Gastric Cancer. *Anticancer Res.* **2018**, *38*, 6827–6833. [CrossRef]
53. Xiong, L.; Liu, Z.; Ouyang, G.; Lin, L.; Huang, H.; Kang, H.; Chen, W.; Miao, X.; Wen, Y. Autophagy inhibition enhances photocytotoxicity of Photosan-II in human colorectal cancer cells. *Oncotarget* **2017**, *8*, 6419–6432. [CrossRef]
54. Ji, H.T.; Chien, L.T.; Lin, Y.H.; Chien, H.F.; Chen, C.T. 5-ALA mediated photodynamic therapy induces autophagic cell death via AMP-activated protein kinase. *Mol. Cancer* **2010**, *9*, 91. [CrossRef]
55. Martins, W.K.; Belotto, R.; Silva, M.N.; Grasso, D.; Suriani, M.D.; Lavor, T.S.; Itri, R.; Baptista, M.S.; Tsubone, T.M. Autophagy Regulation and Photodynamic Therapy: Insights to Improve Outcomes of Cancer Treatment. *Front. Oncol.* **2021**, *10*, 610472. [CrossRef] [PubMed]
56. Zhang, L.; Ji, Z.; Zhang, J.; Yang, S. Photodynamic therapy enhances skin cancer chemotherapy effects through autophagy regulation. *Photodiagn Photodyn. Ther.* **2019**, *28*, 159–165. [CrossRef] [PubMed]
57. Vasiliou, V.; Vasiliou, K.; Nebert, D.W. Human ATP-binding cassette (ABC) transporter family. *Hum. Genom.* **2009**, *3*, 281–290. [CrossRef]
58. Choi, Y.H.; Yu, A.M. ABC transporters in multidrug resistance and pharmacokinetics, and strategies for drug development. *Curr. Pharm. Des.* **2014**, *20*, 793–807. [CrossRef]
59. Saltaji, H.; Armijo-Olivo, S.; Cummings, G.G.; Amin, M.; Da Costa, B.R.; Flores-Mir, C. Influence of blinding on treatment effect size estimate in randomized controlled trials of oral health interventions. *BMC Med. Res. Methodol.* **2018**, *18*, 42. [CrossRef] [PubMed]
60. Gu, S.Y.; Tang, W.P.; Zeng, Y.; Tang, W.P.; Zhao, M.L.; Deng, H.H.; Li, Q. An epithelial cell line established from poorly differentiated nasopharyngeal carcinoma. *Ai Zheng* **1983**, *2*, 70–72.
61. Cheung, S.T.; Huang, D.P.; Hui, A.B.; Lo, K.W.; Ko, C.W.; Tsang, Y.S.; Wong, N.; Whitney, B.M.; Lee, J.C. Nasopharyngeal carcinoma cell line (C666-1) consistently harbouring Epstein-Barr virus. *Int. J. Cancer* **1999**, *83*, 121–126. [CrossRef]
62. Huang, D.P.; Ho, J.H.C.; Poon, Y.F.; Chew, E.C.; Saw, D.; Lui, M.; Li, C.L.; Mak, L.S.; Lai, S.H.; Lau, W.H. Establishment of a cell line (NPC/HK1) from a differentiated squamous carcinoma of the nasopharynx. *Int. J. Cancer* **1980**, *26*, 127–132. [CrossRef] [PubMed]
63. Page, M.J.; McKenzie, J.E.; Bossuyt, P.M.; Boutron, I.; Hoffmann, T.C.; Mulrow, C.D.; Shamseer, L.; Tetzlaff, J.M.; Akl, E.A.; Brennan, S.E.; et al. The PRISMA 2020 statement: An updated guideline for reporting systematic reviews. *BMJ* **2021**, *372*, 71. [CrossRef]

biomedicines MDPI

Article

The Evaluation of SWEEPS Plus Antimicrobial Photodynamic Therapy with Indocyanine Green in Eliminating *Enterococcus faecalis* Biofilm from Infected Root Canals: An In Vitro Study

Golriz Rostami [1,†], Shima Afrasiabi [1,†], Stefano Benedicenti [2], Antonio Signore [3] and Nasim Chiniforush [2,*]

1 Laser Research Center of Dentistry, Dentistry Research Institute, Tehran University of Medical Sciences, Tehran 1441987566, Iran; golriz.rostami@gmail.com (G.R.); shafrasiabi@alumnus.tums.ac.ir (S.A.)
2 Department of Surgical Sciences and Integrated Diagnostics, University of Genoa, Viale Benedetto XV 6, 16132 Genoa, Italy; benedicenti@unige.it
3 Therapeutic Dentistry Department, Institute of Dentistry, I.M. Sechenov First Moscow State Medical University, Trubetskaya Str., 8, b. 2, 119992 Moscow, Russia; dr.signore@icloud.com
* Correspondence: nasimch2002@yahoo.com; Tel.: +98-21-8838-4331
† These authors contributed equally to this work.

Abstract: Objectives: This study aimed to assess the efficacy of shockwave-enhanced emission photoacoustic streaming (SWEEPS) plus antimicrobial photodynamic therapy (aPDT) using indocyanine green (ICG) for the elimination of *Enterococcus faecalis* biofilm from infected root canals. Materials and Methods: thirty sound human single-canal teeth were chosen and standardized to have 12 mm of root length. The root canals were shaped and prepared by means of ProTaper rotary files. After sterilization of the teeth, the canals were inoculated with *E. faecalis* for 2 weeks. The teeth were then randomly divided into six groups (n = five) of control, ICG, ICG + 808 nm diode laser, ICG + SWEEPS, ICG + 808 nm diode laser + SWEEPS, and 5.25% sodium hypochlorite (NaOCl). Following treatment, the number of colony-forming units (CFUs)/mL were calculated for each group. Statistical analysis was carried out using one-way ANOVA. For multiple comparisons, Tukey's test was used as the post hoc test. Results: NaOCl alone showed the highest efficacy ($p < 0.001$). The ICG + 808 nm diode laser + SWEEPS group displayed significantly lower amounts of bacteria than either the ICG + 808 nm diode laser or SWEEPS ($p < 0.001$). There was a statistically significant difference detected between the ICG + 808 nm diode laser and ICG + SWEEPS ($p = 0.035$). Conclusions: SWEEPS can effectively increase the photosensitizer distribution in the root canal space, and its application along with irrigants can bring about promising results.

Keywords: antimicrobial photodynamic therapy; biofilms; disinfection; Er:YAG laser; *Enterococcus faecalis*; indocyanine green; hypochlorite sodium; SWEEPS

1. Introduction

Endodontic treatments aim to effectively reduce the microorganisms responsible for endodontic infections [1]. However, the complete elimination of endodontic pathogens is extremely difficult, if not impossible, with the commonly used instrument methods, due to the complex anatomy of the root canal system and the presence of lateral canals, isthmi, ramifications and fins [2]. *Enterococcus faecalis* is associated with secondary endodontic infections, refractory infected lesions and periapical biofilms, resulting in endodontic treatment failure [3]. Teeth with failed endodontic treatment are more likely than non-endodontically treated teeth to contain this microorganism in their root canal system [3]. The resistance of this bacterium to the challenges of survival within the root canal space is related to the ability to invade the dentinal tubules and bond to collagen fibers, biofilm formation, and its capacity to endure harsh environments [3].

Root canal irrigation is performed along with mechanical cleaning and instrumentation of canals to chemically decrease the intracanal microbial load. Syringe irrigation is the standard method of root canal irrigation. The elimination of bacterial biofilm is not possible merely by the chemical action of irrigants or mechanical instrumentation alone, and chemical irrigants should be used in combination with physical manipulation of the canal in order to be able to access all parts of the root canal system [4]. Instrumentation with rotary and hand files cannot efficiently clean the isthmi and canal irregularities, and approximately 35% of the canal surface always remains intact [5]. In addition, rotary instruments create significant amounts of dentinal debris that may accumulate in the canal irregularities and isthmi. The presence of debris prevents the optimal sealing of the canal with root filling materials, and can impair efficient root canal disinfection [6]. Sodium hypochlorite (NaOCl) is a root canal irrigating solution that is currently the most popular, since it can remove bacteria and their biofilm and dissolve the residual vital and necrotic tissues [7]. Nevertheless, NaOCl has neurotoxic and cytotoxic effects, and exhibits a destructive effect on mineralized dentin [8]. Different techniques are used to enhance the efficacy and penetration depth of irrigants into the canal irregularities, such as sonic and ultrasonic instruments and different types of lasers [9].

Laser application for the activation of root canal irrigants and elimination of debris accumulated in the canal has gained increasing attention in recent years. In antimicrobial photodynamic therapy (aPDT), the root canals are filled with a light-sensitive material known as photosensitizer, which is then activated with the appropriate wavelength of light, and produces singlet oxygen and other free radicals in the presence of oxygen molecules. Free oxygen radicals damage the microbial molecules such as proteins, membrane lipids and nucleic acid, and cause microbial death [10].

Indocyanine green (ICG) (4,5-benzoindotricarbocyanine—C43H47N2NaO6S2), also known as cardio green, is a polymethine dye with 775 kDa molecular weight, and is a water-soluble anionic photosensitizer. Its negative charge decreases its interaction with negatively charged cell membranes. This photosensitizer has a higher absorbance peak (~800 nm) than the conventional photosensitizers [11]. Unlike other photosensitizers, the primary effect of ICG is due to its photothermal, rather than photochemical effects [12]. Thus, it can more effectively excite the electrons and transfer energy for the generation of free radicals. In fact, due to combined photothermal and photochemical effects, ICG is a suitable agent for effective elimination of endodontic pathogens from hard-to-reach and inaccessible areas. This photosensitizer has a simple application, low cytotoxicity, and is quickly excreted from the body [11].

Laser-activated irrigation (LAI) refers to the activation of irrigants with a specific laser wavelength. Lasers used for this purpose include erbium lasers such as the erbium chromium: yttrium-scandium-gallium-garnet (Er,Cr:YSGG) laser with 2780 nm wavelength, and the erbium: yttrium-aluminum-garnet (Er:YAG) laser with 2940 nm wavelength, which are well absorbed in water, and with their mechanism of action based on causing cavitation in irrigating solutions [13,14].

Shockwave-enhanced emission photoacoustic streaming (SWEEPS) is a novel LAI technique suggested for more efficient cleaning of the root canals by using irrigants [15]. In this method, the Er:YAG laser fiber tip is placed in the access cavity filled with irrigant to irradiate the irrigant with paired pulses [16,17]. In this technique, during the collapse of the bubble primarily created by laser irradiation, the second pulse is emitted, creating another bubble, which causes a faster and more violent collapse of the first bubble. The accelerated collapse of the primary bubble, as well as the collapse of the secondary bubble, result in the generation of a shockwave in the irrigant which increases the efficacy of canal disinfection [18]. In other words, the secondary bubble exerts pressure on the primary one and causes its movement into deeper areas and the turbulent movement of the irrigant. For this reason, this method is more efficient than ultrasonic techniques and photon-induced photoacoustic streaming (PIPS) in the elimination of canal debris. In this technique, the determination of the optimal pulse interval is not possible for the clinicians [18]. In auto-SWEEPS mode, which is a more recent technology, this time interval is automatically adjusted between 300–650 μs in 10 μs steps [19]. This study aimed to assess the efficacy of the SWEEPS technique plus aPDT with ICG in eliminating *E. faecalis* biofilm from infected root canals.

2. Materials and Methods

2.1. Sample Preparation

The study protocol was approved by the Ethics Committee of the Tehran University of Medical Sciences (IR. TUMS. DENTISTRY.REC. 1401. 143). Thirty single-rooted teeth with completely formed roots and mature apices that had been extracted for purposes not related to this study were collected. Immediately after extraction, the teeth were cleaned of tissue residues using a brush, and were stored in saline. Next, the teeth were decoronated at the cementoenamel junction using a high-speed handpiece and diamond fissure bur under air and water spray, such that the root length was standardized to be 12 mm. A #15 K-file (Mani Inc., Tochigi, Japan) was introduced into the canal until its tip was visible at the apex. The working length was determined to be 0.5 mm shorter than this length. The canals were then instrumented with the ProTaper rotary system (Dentsply Maillefer, Ballaigues, Switzerland) up to F4 to the working length with the single length technique, as instructed by the manufacturer. In the process of cleaning and shaping, the root canals were irrigated with NaOCl. In addition, 1 mL of 17% ethylenediaminetetraacetic acid (EDTA) (Masterdent, New York, USA) was used for 3 min for smear layer removal, followed by irrigation with 1 mL of saline, NaOCl, for 3 min, and, as the final irrigation step, the canals were rinsed with 5 mL of sterile saline [20]. The root canals were then dried with #40 paper points. To prevent apical leakage through the apex, the apex of the teeth was sealed with auto-polymerizing glass ionomer (GC Gold Label, Kyoto, Japan). To prevent external microbial contamination, the external root surfaces, except for the canal orifice, were coated with one layer of nail varnish. The teeth were then autoclave-sterilized at 121 °C and 15 Psi pressure for 20 min.

2.2. Bacterial Culture

The microorganism used in this study was *E. faecalis* (IBRC-M 11,130), which was obtained from the Iranian Biological Resource Center (Tehran, Iran). *E. faecalis* was cultured in brain heart infusion broth (Ibresco, Iran) under aerobic conditions at 37 °C, overnight. Bacterial suspension with 0.5 McFarland standard concentration (1.5×10^8 colony-forming unit (CFU)/mL) was prepared using a spectrophotometer (optical density (OD) 600 nm: 0.08–0.13). After sterilization of the teeth, the root canals were inoculated with 10 µL of *E. faecalis* bacterial suspension (1.5×10^7 CFU/mL) using a micropipette, and the teeth were incubated at 37 °C for 2 weeks. Ten microliters of fresh microbial suspension were inoculated into the canals every 48 h. After termination of the incubation period, the teeth were rinsed with sterile saline, and randomly assigned to 6 groups.

2.3. Scanning Electron Microscope (SEM) Measurements

After performing the above steps, in order to confirm the formation of biofilm, the teeth were sectioned vertically into two parts and fixed in 1% aqueous osmium tetroxide followed by an ethanol gradient wash, and then sputter coated with gold. The samples were imaged with a SEM-EDAX apparatus (FEI SEM QUANTA 200 EDAX EDS SILICON DRIFT 2017, Hillsborough, OR, USA) at a magnification of 3000×.

2.4. Study Groups

The treatment steps were as follows (Figure 1):

Group 1. Control group: the teeth did not undergo any intervention.

Group 2. ICG: the root canals were filled with 10 µL of ICG (Green + I, NovaTeb Pars, Tehran, Iran) at a concentration of 1000 µg/mL, and placed at room temperature in the dark for 5 min.

Group 3. ICG + 808 nm diode laser: the root canals were filled with 10 µL of ICG (1000 µg/mL) and placed at room temperature in the dark for 5 min. They were then subjected to 808 nm diode laser (DX82, Konftec, New Taipei City, Taiwan) with output power of 250 mW and total energy of 15 J, for 60 s. The 3D diffuser tip was used in an up-and-down motion from the apex to the coronal part.

Group 4. ICG + SWEEPS: the root canals were filled with 10 μL of ICG (1000 μg/mL) and placed at room temperature in the dark for 5 min. They were then subjected to Er:YAG laser irradiation with 2940 nm wavelength (LightWalker AT, Fotona, LjuBlijana, Slovenia) with an H14 handpiece and SWEEPS tip with the Fotona protocol for SWEEPS (25 μs, SWEEPS mode, 15 Hz, 20 mJ, 0.3 W). The tip of SWEEPS was placed in the pulp chamber and activated for 90 s.

Group 5. ICG + 808 nm diode laser + SWEEPS: the root canals were treated by ICG-mediated SWEEPS similar to group 4, and then, after 5 min, 808-nm diode laser irradiation was performed, similar to group 3.

Group 6. NaOCl: the root canals were filled with 5.25% NaOCl for 1 min.

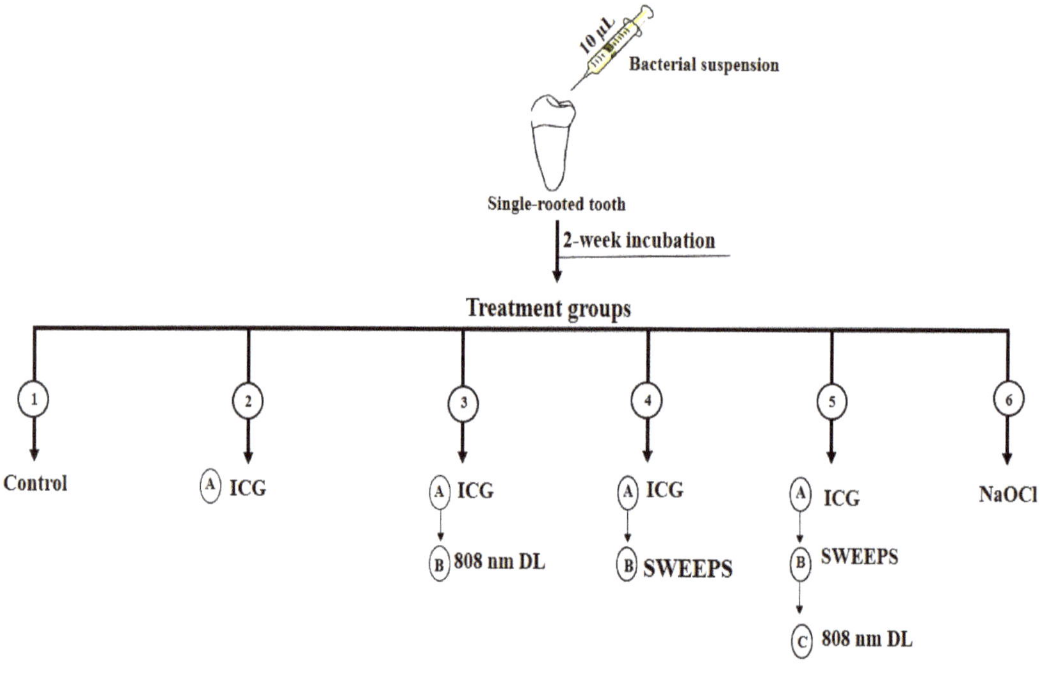

Figure 1. Schematic representation of experimental setup. ICG: Indocyanine green, nm: nanometer, DL: diode laser, SWEEPS: shockwave-enhanced emission photoacoustic streaming, NaOCl: 5.25% sodium hypochlorite.

2.5. *Microbiological Process*

After treatment, the teeth were placed in a microtube containing 1 mL BHI broth, and vortexed for 1 min. Next, 10 μL of the suspension was serially diluted 5 times, and 10 μL of each dilution was cultured on BHI agar (Ibresco), and incubated at 37 °C for 24 h. The colonies were then counted [11].

2.6. *Statistical Analysis*

Statistical analysis was carried out using one-way ANOVA (SPSS, version 23.0, Chicago, IL, USA). For multiple comparisons, Tukey's test was used as the post hoc test. p value < 0.05 was considered statistically significant.

3. Results

The SEM of the root canal without bacterial biofilm and *E. faecalis* biofilm on the root canal walls and in the dentinal tubules 2 weeks after inoculation are shown in Figure 2A,B, respectively. The results in Figure 3 and Table 1 demonstrate that, except for ICG alone, all experimental groups could decrease the viability of *E. faecalis*, compared with the control ($p < 0.001$). The results revealed that NaOCl decreased the microbial count to almost zero ($p < 0.001$). A significant difference was found between the ICG and ICG + SWEEPS or 808 nm diode laser, or both ($p < 0.001$) regarding the reduction in the *E. faecalis* count. Accordingly, ICG + 808 nm diode laser + SWEEPS had a 2.1- and 1.3-fold anti-biofilm effect compared to ICG + SWEEPS and ICG + 808 nm diode laser, respectively. In addition, there was a significant difference between the ICG + SWEEPS and ICG + 808 nm diode laser + SWEEPS ($p = 0.002$), whereas there was no significant difference found between the ICG + 808 nm diode laser and ICG + 808 nm diode laser + SWEEPS ($p = 0.64$). Furthermore, ICG + SWEEPS and ICG + 808 nm diode laser groups showed significant differences ($p = 0.035$).

Table 1. Multiple comparison post hoc test results.

Groups	Groups	*p* Value
ICG	Control	0.86
	ICG + DL	<0.001
	ICG + SWEEPS	<0.001
	ICG + DL+ SWEEPS	<0.001
	NaOCl	<0.001
ICG + DL	Control	<0.001
	ICG + SWEEPS	0.035
	ICG + DL+ SWEEPS	0.64
	NaOCl	0.001
ICG + SWEEPS	Control	<0.001
	ICG + DL + SWEEPS	0.002
	NaOCl	<0.001
ICG + DL + SWEEPS	Control	<0.001
	NaOCl	0.01
NaOCl	Control	<0.001

ICG: indocyanine green, DL: 808 nm diode laser, SWEEPS: shockwave-enhanced emission photoacoustic streaming, NaOCl: 5.25% sodium hypochlorite.

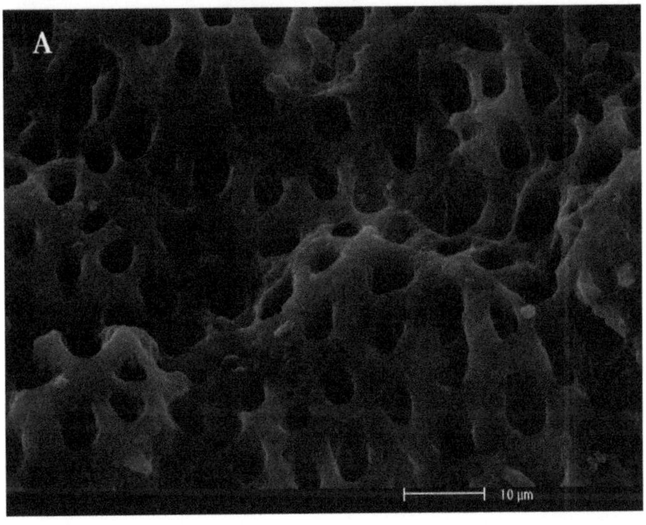

Figure 2. *Cont.*

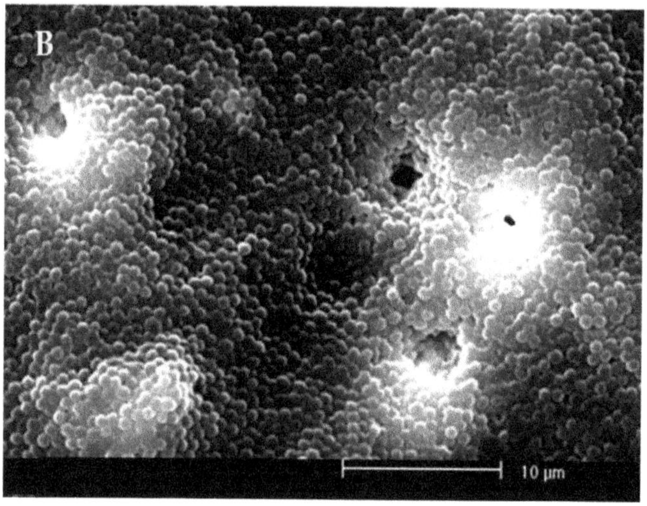

Figure 2. (**A**) Scanning electron microscope of the root canal without bacterial biofilm. (**B**) Scanning electron microscope of *Enterococcus faecalis* biofilm on the root canal walls at a magnification of 3000×.

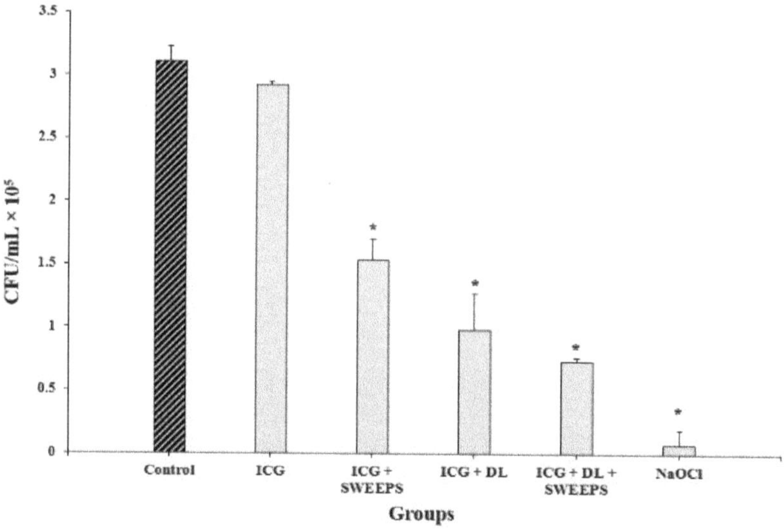

Figure 3. Effect of different treatment groups on cell viability of *Enterococcus faecalis* biofilm. * Significantly different from the control group, $p < 0.001$. ICG: indocyanine green, DL: 808 nm diode laser, SWEEPS: shockwave-enhanced emission photoacoustic streaming, NaOCl: 5.25% sodium hypochlorite.

4. Discussion

This study evaluated the efficacy of SWEEPS plus aPDT with ICG for the reduction of *E. faecalis* biofilm from the root canal space. Since studies on the efficacy of the SWEEPS technique in actual root canals are highly limited, this study assessed the efficacy of the above-mentioned techniques in actual root canals with the usual anatomical complexities.

The present results revealed that the application of NaOCl alone had a significantly superior efficacy than the other groups in relation to the reduction of the *E. faecalis* count in the root canal system. For higher disinfecting efficacy, antimicrobial agents need to be in direct contact with the bacteria; however, the lodging of bacteria in anatomical complexities of the canal such as anastomoses, fins, and ramifications make it impossible for the irrigants to directly contact the microorganisms [20].

In the present study, aPDT with 808 nm diode laser and ICG caused a significant reduction in the *E faecalis* count, which is in consistent with the results of most studies conducted on the efficacy of aPDT with ICG on *E. faecalis* elimination [21,22]. A study has also shown that there is no evidence of cytotoxicity of ICG to MG-63 human osteoblast-like cells [23]. Furthermore, the results of this study show that ICG + SWEEPS has significant efficacy in removing *E. faecalis* biofilms compared to the control ($p < 0.001$). Wang et al. [24] confirmed the efficiency of the auto-SWEEPS in eliminating *E. faecalis* biofilm in root canals compared to 3% NaOCl alone and PIPS, using SEM images. They explained that strong shockwaves are eventually generated throughout the root canal, which significantly improves clearance efficacy.

In the SWEEPS technique, a photothermal effect does not occur, due to subablative laser irradiation. This technique generates powerful waves in the irrigating liquid, and produces a high fluid flow rate [25]. The maximum speed of the irrigating solution in accessory canals in the application of the SWEEPS technique is 10 m/s, which is much higher than the reported speed for other methods of root canal irrigation [26]. In addition, the penetration depth of the irrigant into the accessory canals in the SWEEPS technique is more than 1 mm [26]. According to the literature, the optimal efficacy of the SWEEPS technique is attributed to the emission of two pulses with an optimal time interval. Accordingly, the effect of the primary bubbles is reinforced by the generation of secondary bubbles, without increasing the risk of extrusion of the irrigant. Thus, this technique increases the efficacy of irrigation and debris removal from the root canal system [27]. Su et al. [26] described the breath mode for the streaming of irrigants, in which the irrigant repeatedly enters and exits the main and accessory canals. According to their study, another advantage of the SWEEPS technique is that the in-and-out movement of the liquid, which resembles inhalation and exhalation, generates alternating shear stresses in the root canal, which plays a pivotal role in improving the quality of the debridement process. No risk of intracanal instrument fracture is another advantage of the SWEEPS technique, since in this technique the tip of the handpiece is positioned in the access cavity and above the root canals, whereas in sonic and ultrasonic techniques, the instrument is introduced into the canal and proceeds to the apex for the activation of the intracanal irrgant [28]. In addition, a previous study has shown that photosensitizer extrusion is not harmful [29].

According to the present results, combination therapy with the application of SWEEPS plus aPDT yielded superior results. The SWEEPS technique mechanically detaches the biofilm from the root canal, therefore increasing the penetration depth and efficacy of the photosensitizer, due to the frequent generation of cavitation [1]. Consistent with the present results, previous studies also showed that application of SWEEPS with a 660 and 980 nm diode laser or light-emitting diode (LED) can cause a more significant decrease in the root canal infected with *E. faecalis* [25,30]. However, in contrast with the results of this study, the diode laser and SWEEPS did not show significant differences with methylene blue in their use as a photosensitizer [30]. Since the type of photosensitizer, light source, and irradiation time affect the bactericidal properties, this difference might be related to the fact that the mechanism of action of ICG is different from that of other photosensitizers. The main effect of ICG is due to the photothermal effect, which causes cell damage by increasing the intracellular temperature [12]. In photothermal therapy, the energy of laser radiation is absorbed by ICG, effectively raising the local temperature [31]. In addition to photothermic effects, ICG was demonstrated to have a photodynamic effect via the production of reactive oxygen species [32]. On the other hand, the 810 nm diode laser compared to other wavelengths used for toluidine blue O and methylene blue allows more penetration depth [12]. The application of only one type of photosensitizer was among the limitations of this study. The in vitro design was another limitation of this study that limits the clinical generalizability of the results. Therefore, future studies are required on other types of photosensitizers and irrigants, and also in the clinical setting.

5. Conclusions

The application of ICG along with the SWEEPS plus aPDT significantly improve its efficacy compared with its application alone. The results of this study will probably make an important contribution in the future to improving the efficiency of root canal treatments.

Author Contributions: Conceptualization, S.A., S.B., A.S. and N.C.; methodology, S.A. and N.C.; software, S.A.; writing—original draft preparation, G.R. and S.A.; writing—review and editing, S.A. All authors have read and agreed to the published version of the manuscript.

Funding: This research received no external funding.

Institutional Review Board Statement: Not applicable.

Informed Consent Statement: Not applicable.

Data Availability Statement: Not applicable.

Conflicts of Interest: The authors declare no conflict of interest.

References

1. Ivanusic, T.; Lukac, M.; Lukac, N.; Jezersek, M. SSP/SWEEPS endodontics with the SkyPulse Er: YAG dental laser. *J. LAHA* **2019**, *2019*, 1–10.
2. Miranda, T.C.; Andrade, J.F.M.; Gelfuso, G.M.; Cunha-Filho, M.; Oliveira, L.A.; Gratieri, T. Novel technologies to improve the treatment of endodontic microbial infections: Inputs from a drug delivery perspective. *Int. J. Pharm.* **2023**, *635*, 122794. [CrossRef]
3. Stuart, C.H.; Schwartz, S.A.; Beeson, T.J.; Owatz, C.B. *Enterococcus faecalis*: Its role in root canal treatment failure and current concepts in retreatment. *J. Endod.* **2006**, *32*, 93–98. [CrossRef]
4. Nagahashi, T.; Yahata, Y.; Handa, K.; Nakano, M.; Suzuki, S.; Kakiuchi, Y.; Tanaka, T.; Kanehira, M.; Suresh Venkataiah, V.; Saito, M. Er: YAG laser-induced cavitation can activate irrigation for the removal of intraradicular biofilm. *Sci. Rep.* **2022**, *12*, 4897. [CrossRef]
5. Peters, O.A.; Schönenberger, K.; Laib, A. Effects of four Ni-Ti preparation techniques on root canal geometry assessed by micro computed tomography. *Int. Endod. J.* **2001**, *34*, 221–230. [CrossRef]
6. Yang, Q.; Liu, M.; Zhu, L.; Peng, B. Micro-CT study on the removal of accumulated hard-tissue debris from the root canal system of mandibular molars when using a novel laser-activated irrigation approach. *Int. Endod. J.* **2020**, *53*, 529–538. [CrossRef] [PubMed]
7. Oliveira, L.; Carvalho, C.A.T.; Nunes, W.; Valera, M.C.; Camargo, C.H.R.; Jorge, A.O.C. Effects of chlorhexidine and sodium hypochlorite on the microhardness of root canal dentin. *Oral. Surg. Oral. Med. Oral. Pathol. Oral. Radiol. Endodontol.* **2007**, *104*, 125–128. [CrossRef]
8. Paiva, S.S.M.; Siqueira Jr, J.F.; Rôças, I.N.; Carmo, F.L.; Leite, D.C.A.; Ferreira, D.C.; Rachid, C.T.C.; Rosado, A.S. Clinical antimicrobial efficacy of NiTi rotary instrumentation with NaOCl irrigation, final rinse with chlorhexidine and interappointment medication: A molecular study. *Int. Endod. J.* **2012**, *46*, 225–233. [CrossRef]
9. Dioguardi, M.; Di Gioia, G.; Illuzzi, G.; Laneve, E.; Cocco, A.; Troiano, G. Endodontic irrigants: Different methods to improve efficacy and related problems. *Eur. J. Dent.* **2018**, *12*, 459. [CrossRef]
10. Cieplik, F.; Deng, D.; Crielaard, W.; Buchalla, W.; Hellwig, E.; Al-Ahmad, A.; Maisch, T. Antimicrobial photodynamic therapy—What we know and what we don't. *Crit. Rev. Microbiol.* **2018**, *44*, 571–589. [CrossRef]
11. Beltes, C.; Economides, N.; Sakkas, H.; Papadopoulou, C.; Lambrianidis, T. Evaluation of antimicrobial photodynamic therapy using indocyanine green and near-infrared diode laser against *Enterococcus faecalis* in infected human root canals. *Photomed. Laser Surg.* **2017**, *35*, 264–269. [CrossRef]
12. Monzavi, A.; Chinipardaz, Z.; Mousavi, M.; Fekrazad, R.; Moslemi, N.; Azaripour, A.; Bagherpasand, O.; Chiniforush, N. Antimicrobial photodynamic therapy using diode laser activated indocyanine green as an adjunct in the treatment of chronic periodontitis: A randomized clinical trial. *Photodiagn. Photodyn. Ther.* **2016**, *14*, 93–97. [CrossRef]
13. Kuştarcı, A.; Er, K. Efficacy of laser activated irrigation on apically extruded debris with different preparation systems. *Photomed. Laser Surg.* **2015**, *33*, 384–389. [CrossRef]
14. Lukač, M.; Olivi, G.; Constantin, M.; Lukač, N.; Jezeršek, M. Determination of Optimal Separation Times for Dual-Pulse SWEEPS Laser-Assisted Irrigation in Different Endodontic Access Cavities. *Lasers Surg. Med.* **2021**, *53*, 998–1004. [CrossRef]
15. Lei, L.; Wang, F.; Wang, Y.; Li, Y.; Huang, X. Laser Activated Irrigation with SWEEPS Modality Reduces Concentration of Sodium Hypochlorite in Root Canal Irrigation. *Photodiagnosis. Photodyn. Ther.* **2022**, *39*, 102873. [CrossRef] [PubMed]
16. Panthangi, S.; Vishwaja, U.; Reddy, C.L.C.; Babu, M.B.; Podili, S. Novel sweeps technology in endodontics—A review. *IP Indian J. Conserv. Endod.* **2021**, *6*, 134–142.

17. Wang, M.; Ma, L.; Li, Q.; Yang, W. Efficacy of Er: YAG laser-assisted direct pulp capping in permanent teeth with cariously exposed pulp: A pilot study. *Aust. Endod. J.* **2020**, *46*, 351–357. [CrossRef]
18. Lukač, M.; Lukač, N.; Jezeršek, M. Characteristics of bubble oscillations during laser-activated irrigation of root canals and method of improvement. *Lasers Surg. Med.* **2020**, *52*, 907–915. [CrossRef]
19. Lukac, N.; Muc, B.T.; Jezersek, M.; Lukac, M. Photoacoustic endodontics using the novel SWEEPS Er: YAG laser modality. *J. Laser Health Acad.* **2017**, *1*, 1–7.
20. Balić, M.; Lucić, R.; Mehadžić, K.; Bago, I.; Anić, I.; Jakovljević, S.; Plečko, V. The efficacy of photon-initiated photoacoustic streaming and sonic-activated irrigation combined with QMiX solution or sodium hypochlorite against intracanal *E. faecalis* biofilm. *Lasers Med. Sci.* **2016**, *31*, 335–342. [CrossRef]
21. Beltes, C.; Sakkas, H.; Economides, N.; Papadopoulou, C. Antimicrobial photodynamic therapy using Indocyanine green and near-infrared diode laser in reducing *Entrerococcus faecalis*. *Photodiagn. Photodyn. Ther.* **2017**, *17*, 5–8. [CrossRef]
22. Chiniforush, N.; Pourhajibagher, M.; Parker, S.; Shahabi, S.; Bahador, A. The in vitro effect of antimicrobial photodynamic therapy with indocyanine green on *Enterococcus faecalis*: Influence of a washing vs. non-washing procedure. *Photodiagn. Photodyn. Ther.* **2016**, *16*, 119–123. [CrossRef]
23. Aghayan, S.; Yazdanfar, A.; Seyedjafari, E.; Noroozian, M.; Ioana Bordea, R.; Chiniforush, N. Evaluation of indocyanine-mediated photodynamic therapy cytotoxicity in human osteoblast-like cells: An in vitro study. *Folia Med.* **2022**, *64*, 932–937. [CrossRef] [PubMed]
24. Wang, X.-N.; Shi, J. Shock wave-enhanced emission photoacoustic streaming versus photon-induced photoacoustic streaming modes for clearing root canal bacteria using erbium-doped yttrium aluminum garnet lasers: An in vitro study. *Res. Sq.* **2020**. [CrossRef]
25. Ensafi, F.; Fazlyab, M.; Chiniforush, N.; Akhavan, H. Comparative effects of SWEEPS technique and antimicrobial photodynamic therapy by using curcumin and nano-curcumin on *Enterococcus faecalis* biofilm in root canal treatment. *Photodiagn. Photodyn. Ther.* **2022**, *40*, 103130. [CrossRef]
26. Su, Z.; Li, Z.; Shen, Y.; Bai, Y.; Zheng, Y.; Pan, C.; Hou, B. Characteristics of the Irrigant Flow in a Simulated Lateral Canal Under Two Typical Laser-Activated Irrigation Regimens. *Lasers Surg. Med.* **2021**, *53*, 587–594. [CrossRef]
27. Jezeršek, M.; Lukač, N.; Lukač, M.; Tenyi, A.; Olivi, G.; Fidler, A. Measurement of pressures generated in root canal during Er: YAG laser-activated irrigation. *Photobiomodul. Photomed. Laser Surg.* **2020**, *38*, 625–631. [CrossRef]
28. Widbiller, M.; Keim, L.; Schlichting, R.; Striegl, B.; Hiller, K.-A.; Jungbauer, R.; Buchalla, W.; Galler, K.M. Debris removal by activation of endodontic irrigants in complex root canal systems: A standardized in-vitro-study. *Appl. Sci.* **2021**, *11*, 7331. [CrossRef]
29. Bolhari, B.; Meraji, N.; Seddighi, R.; Ebrahimi, N.; Chiniforush, N. Effect of SWEEPS and PIPS techniques on dye extrusion in photodynamic therapy procedure after root canal preparation. *Photodiagn. Photodyn. Ther.* **2023**, *42*, 103345. [CrossRef]
30. Afrasiabi, S.; Parker, S.; Chiniforush, N. Synergistic antimicrobial effect of photodynamic inactivation and SWEEPS in combined treatment against *Enterococcus faecalis* in a root canal biofilm model: An in vitro study. *Appl. Sci.* **2023**, *13*, 5668. [CrossRef]
31. Sukumar, K.; Tadepalli, A.; Parthasarathy, H.; Ponnaiyan, D. Evaluation of combined efficacy of photodynamic therapy using indocyanine green photosensitizer and non-surgical periodontal therapy on clinical and microbial parameters in the management of chronic periodontitis subjects: A randomized split-mouth design. *Photodiagn. Photodyn. Ther.* **2020**, *31*, 101949. [CrossRef] [PubMed]
32. You, Q.; Sun, Q.; Wang, J.; Tan, X.; Pang, X.; Liu, L.; Yu, M.; Tan, F.; Li, N. A single-light triggered and dual-imaging guided multifunctional platform for combined photothermal and photodynamic therapy based on TD-controlled and ICG-loaded CuS@mSiO_2. *Nanoscale.* **2017**, *9*, 3784–3796. [CrossRef] [PubMed]

biomedicines

MDPI

Article

The Effect of Different Output Powers of Blue Diode Laser along with Curcumin and Riboflavin against *Streptococcus mutans* around Orthodontic Brackets: An In Vitro Study

Edris Pordel [1], Trife Ghasemi [2], Shima Afrasiabi [3,*], Stefano Benedicenti [4], Antonio Signore [5] and Nasim Chiniforush [4,*]

1 Department of Pediatric Dentistry, Dental School, Sabzevar University of Medical Sciences, Sabzevar 9613875389, Iran; edrispordel@yahoo.com
2 Independent Researcher, Mashhad 9613875389, Iran; trifetrife@yahoo.com
3 Laser Research Center of Dentistry, Dentistry Research Institute, Tehran University of Medical Sciences, Tehran 1441987566, Iran
4 Department of Surgical Sciences and Integrated Diagnostics, University of Genoa, Viale Benedetto XV, 6, 16132 Genoa, Italy; benedicenti@unige.it
5 Therapeutic Dentistry Department, Institute of Dentistry, I.M. Sechenov First Moscow State Medical University, Trubetskaya Str. 8, b. 2, 119992 Moscow, Russia; dr.signore@icloud.com
* Correspondence: shafrasiabi@alumnus.tums.ac.ir (S.A.); nasimch2002@yahoo.com (N.C.); Tel.: +98-21-8838-4331 (S.A.); +39-3497374134 (N.C.)

Abstract: Objectives: The aim of the present study was to determine the effects of antimicrobial photodynamic therapy (aPDT) using the blue diode laser (BDL) with different output powers and the photosensitizers riboflavin and curcumin on reducing the number of *Streptococcus mutans* around orthodontic brackets. Materials and methods: A total of 36 orthodontic brackets were contaminated with *S. mutans* and randomly assigned to 12 groups as follows: control, riboflavin alone, riboflavin + BDL with an output power of 200, 300, 400, or 500 mW, and curcumin alone, curcumin + BDL with an output power of 200, 300, 400, or 500 mW, and 0.2% chlorhexidine (CHX-positive control). Orthodontic brackets were irradiated with a BDL (wavelength 445 nm) at a power density of 0.4–1.0 W/cm^2 for 30 s. All orthodontic brackets were examined under a stereomicroscope at 10× magnification. Mean colony-forming units (CFUs)/mL were measured before and after treatment. A one-way analysis of variance with Tukey's post hoc test was performed to compare CFU/mL between groups. Results: CHX and curcumin plus BDL with an output power of 500 mW had the highest reduction in *S. mutans* colony numbers ($p < 0.001$). The curcumin groups were more effective than the riboflavin groups. Riboflavin alone and riboflavin + BDL with an output power of 200 mW showed no significant difference from the control group ($p = 0.99$ and 0.74, respectively). Conclusion: Our results suggest that aPDT using curcumin as a photosensitizer plus BDL with an output power of 500 mW and a power density of 1.0 W/cm^2 at a wavelength of 445 nm can effectively reduce colonies of *S. mutans* around stainless steel brackets.

Keywords: antimicrobial photodynamic therapy; blue diode laser; curcumin; riboflavin; orthodontic brackets; dental caries

1. Introduction

The goal of orthodontic treatment for misaligned teeth is to improve the appearance of the mouth and jaw region and to make chewing easier by realigning the teeth. The side effects of orthodontic therapy include periodontal disease, root resorption, temporomandibular joint disorders, dental caries, and enamel degradation [1]. Dental caries is a severe, contagious, bacterial infection with complex pathophysiology. Cariogenic bacteria, fermentable carbohydrates, a susceptible tooth, the host, and time are the main factors contributing to the development of this disease [2]. It is a major economic burden and public health problem throughout the world [3]. *Streptococcus mutans* is the main cause of dental caries due to its ability to synthesize glucans, form biofilms, and produce acid. Treatment methods commonly used for caries prevention include mechanical devices, antibacterial agents, fluoride therapy, and immunizations [4,5]. Antimicrobial photodynamic therapy (aPDT) is a strategy to suppress cariogenic bacteria and prevent periodontal disease. This procedure is safe in the long term as it has no genotoxic and mutagenic consequences. The advantages of this technique include rapid reduction in bacteria, no invasion and damage to surrounding tissues, accessibility to sites with complicated architecture, low risk of bacteremia, and excellent repeatability [6]. The aPDT method includes a photosensitizer, oxygen, and light. In the presence of oxygen, the photosensitizer generates reactive oxygen species (ROS) and other free radicals upon exposure to light, which lead to irreversible destruction of cellular components, alter signal transduction pathways, and disrupt metabolic processes, ultimately resulting in cell death [7].

Recently, natural photosensitizers such as riboflavin and curcumin have come into focus. Riboflavin, a water-soluble vitamin, and curcumin possess biocompatibility, nontoxicity, and the ability to produce ROS that have recently been investigated for their potential antibacterial effects [8]. Riboflavin is an efficient photosensitizer that causes oxidative damage when activated by visible light, especially blue light [6]. Kamran et al. used a blue LED with a light intensity of 2000 mW/cm^2 and a wavelength range of 450 nm and fluence of 95 J/cm^2 to investigate the antimicrobial potential of riboflavin-mediated aPDT. They showed that the method significantly reduced the amount of *S. mutans* around metallic brackets [9]. Comeau et al. showed that riboflavin with blue LED groups caused a lower number of *S. mutans* colony-forming unit (CFU)/mL compared to the control group [10]. The highest absorption range for riboflavin is at wavelengths of 445, 336, and 270 nm [11].

For curcumin, it is around 300–500 nm with maximum absorption at a wavelength of 425 nm [12]. Curcumin is a hydrophobic photosensitizer that is soluble in polar solvents such as dimethyl sulfoxide (DMSO), acetonitrile, methanol, oils, etc. [13]. Curcumin adheres to the lipid membrane and bacterial proteins and damages the bacterial membrane by various mechanisms. Due to its ability to absorb blue light and generate ROS, it has shown significant potential as a photosensitizer [14]. Previously, Azizi et al. demonstrated in an experimental in vitro study that curcumin-mediated aPDT using 460 nm/100 mW laser (60 s) can considerably reduce *S. mutans* colonies [15]. Lee et al. used a 405 nm light-emitting diode (LED) with a power density of 84.5 mW for 5 min at an energy density of 25.3 J/cm^2 and curcumin as a photosensitizer. They showed that under the above conditions, the viability of *S. mutans* was significantly reduced [16]. Tonon et al. showed that the group irradiated with blue LED in a power density of 240.1 mW/cm^2 and in the presence of curcumin achieved a significantly lower number of viable *S. mutans* compared to the control group [17]. The aim of the present study was to determine the effects of different output powers of blue diode laser (BDL) with the photosensitizers riboflavin and curcumin on reducing the number of *S. mutans* around orthodontic brackets. The null hypothesis was tested that there is no difference between different output powers of BDL during aPDT against *S. mutans* around orthodontic brackets.

2. Materials and Methods

2.1. Sample Size

According to the conditions of $\alpha = 0.05$, $\beta = 0.2$, standard deviation = 8.66, and effect size = 0.45, the sample size was calculated using the one-way analysis of variance (ANOVA) in IBM SPSS Statistics 25.0 (Armonk, NY, USA) and was set to 3 orthodontic brackets in each group.

2.2. Sample Preparation

The study protocol was approved by the Ethics Committee of the Tehran University of Medical Sciences (IR. TUMS. DENTISTRY.REC. 1400. 187). A total of 36 healthy human premolar teeth with no carious were bonded to 0.022, stainless steel brackets (TSHdental, Tehran, Iran). The teeth were extracted for orthodontic reasons. A stereomicroscope (SMZ800, Nikon, Tokyo, Japan) was used to examine the orthodontic brackets at a magnification of 10×. The orthodontic brackets displayed normal architecture and healthy enamel on the buccal surface, free of any cracks, fractures, or restorations in the enamel. No pretreatments for bleaching or aPDT were recorded in the patient records. A periodontal scaler was used to remove the remaining soft tissue around the orthodontic brackets, and all dentin surfaces were polished for 15 s with a low-speed handpiece (Coxo, Guangzhou, China), rubber cups (Microdont, São Paulo, Brazil), and pumice paste (Cina, Tehran, Iran) before being thoroughly washed under tap water. Orthodontic brackets were cleaned with a 0.5% (w/v) chloramine T (MF aqua, Tehran, Iran) solution for one week at 3 °C and then stored in saline until the start of the experiment.

2.3. Microbial Suspension

S. mutans (IBRC-M 10,682) was provided by the Iranian Biological Resource Centre (Tehran, Iran) for this study. Bacteria was inoculated in 10 mL of brain heart infusion (BHI) broth (Ibresco, Tehran, Iran) and were incubated under an aerobic atmosphere at 37 °C overnight. A 1.5×10^8 CFU/mL bacterial suspension equivalent to 0.5 McFarland was prepared. A spectrophotometer (Spectroshade Micro, MHT, Verona, Italy) was used for verification, indicating a value between 0.8 and 0.13 at a wavelength of 625 nm. Orthodontic brackets were placed in a 24-well microplate and contaminated with 1 mL of *S. mutans* suspension (10^6 CFU/mL). Subsequently, samples were incubated at 37 °C for 72 h for biofilm formation.

2.4. Experimental Groups

The current study included 12 groups with three orthodontic brackets, including:

1. Control group;
2. Riboflavin alone;
3. Riboflavin + BDL (output power = 200 mW);
4. Riboflavin + BDL (output power = 300 mW);
5. Riboflavin + BDL (output power = 400 mW);
6. Riboflavin + BDL (output power = 500 mW);
7. Curcumin alone;
8. Curcumin + BDL (output power = 200 mW);
9. Curcumin + BDL (output power = 300 mW);
10. Curcumin + BDL (output power = 400 mW);
11. Curcumin + BDL (output power = 500 mW);
12. 0.2% chlorhexidine mouthwash (CHX; Vi-One, Tabriz, Iran) as positive control.

2.5. Photosensitizers, Light Source, and aPDT

Curcumin (UltraCur, weber medical, Lauenförde, Germany) was diluted with <1% DMSO (Sigma-Aldrich, St. Louis, MO, USA) in phosphate-buffered solution to reach a final concentration of 40 μM. Riboflavin (Harman Finochem Ltd., Mumbai, India) was dissolved in 0.9% NaCl to reach a final concentration of 100 μM. Before starting the experiment, both photosensitizers were stored in a dark room. One milliliter of the riboflavin or curcumin solution was applied to the contaminated orthodontic brackets. The samples were illuminated at room temperature using a BDL (Wiser 3, Doctor Smile, Brendola, Italy) with a wavelength of 445 nm and output powers of 200, 300, 400, and 500 mW for 30 s. The tip diameter was 8 mm, the surface area was considered as 0.5 cm^2, and the distance between the tip and the samples was 1 mm. After treatment, each orthodontic bracket was inserted in microtube containing 1 mL BHI broth and vortexed for 2 min. Then, 10 μL from each solution was diluted (10^{-1}–10^{-5}) and transferred to BHI agar (Ibresco, Tehran, Iran) plates. The plates were incubated under aerobic incubation at 37 °C for 48 h. The number of CFU/mL was determined according to the previously described method of Miles and Misra [18].

2.6. Statistical Analysis

IBM SPSS 25.0 statistics was used for all calculations. Descriptive statistics were obtained for each group, including mean, standard deviation, and minimum and maximum values. Mean and $\log_{10}$ CFU/mL was compared between groups using the ANOVA and a post hoc Tukey's test. The significance threshold was set at $p < 0.05$.

3. Results

As shown in Figure 1, study groups with higher output power caused less viable bacteria around orthodontic brackets. The details of the analysis between groups and the results of Tukey's tests are shown in Table 1. Post hoc analysis of the $\log_{10}$ CFU/mL showed that the riboflavin + BDL with an output power of 200 or 300 mW has a significant difference from the riboflavin + BDL with an output power of 500 mW ($p < 0.001$). Likewise, a significant difference was observed between curcumin + BDL with an output power of 200 mW and curcumin + BDL with an output power of 400 or 500 mW ($p = 0.01$ and <0.001, respectively). However, no significant difference was found between BDL with an output power of 200 mW and 300 mW, as well as BDL with an output power of 400 mW and 500 mW, in any of the photosensitizer groups. Moreover, only riboflavin alone and riboflavin + BDL with an output power of 200 mW failed to significantly eliminate the amount of bacteria around orthodontic brackets. In addition, all curcumin groups were significantly more capable of reducing bacteria around orthodontic brackets than riboflavin groups.

Table 1. Comparison between groups based on mean colony count of *Streptococcus mutans* after intervention.

Group 1	Group 2	*p*-Value
Riboflavin	Riboflavin + BDL 200 mW	0.85
	Riboflavin + BDL 300 mW	0.03 *
	Riboflavin + BDL 400 mW	0.00 *
	Riboflavin + BDL 500 mW	0.00 *
	CHX	0.00 *
Riboflavin + BDL 200 mW	Riboflavin + BDL 300 mW	0.67
	Riboflavin + BDL 400 mW	0.07
	Riboflavin + BDL 500 mW	0.00 *
	CHX	0.00 *
Riboflavin + BDL 300 mW	Riboflavin + BDL 400 mW	0.95
	Riboflavin + BDL 500 mW	0.00 *
	CHX	0.00 *
Riboflavin + BDL 400 mW	Riboflavin + BDL 500 mW	0.06
	CHX	0.00 *
Riboflavin + BDL 500 mW	CHX	0.00 *
Curcumin	Curcumin + BDL 200 mW	0.00 *
	Curcumin + BDL 300 mW	0.00 *
	Curcumin + BDL 400 mW	0.00 *
	Curcumin + BDL 500 mW	0.00 *
	CHX *	0.00 *
Curcumin + BDL 200 mW	Curcumin + BDL 300 mW	0.10
	Curcumin + BDL 400 mW	0.01 *
	Curcumin + BDL 500 mW	0.00 *
	CHX	0.00 *
Curcumin + BDL 300 mW	Curcumin + BDL 400 mW	0.99
	Curcumin + BDL 500 mW	0.87
	CHX	0.89
Curcumin + BDL 400 mW	Curcumin + BDL 500 mW	1.00
	CHX	1.00
Curcumin + BDL 500 mW	CHX	1.00

Abbreviation: BDL: blue diode laser, CXH: chlorhexidine. * Significantly different compared to other group.

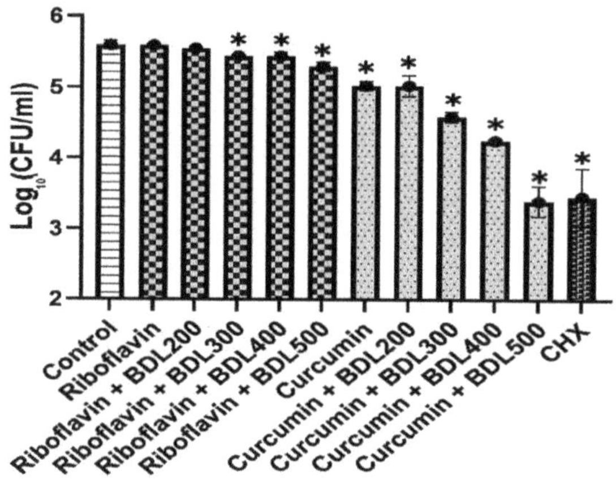

Figure 1. Log_{10} CFU/mL of *Streptococcus mutans* of the study groups. * Significantly different compared to control group. BDL: blue diode laser; CHX: chlorhexidine.

4. Discussion

Orthodontic brackets and ligature wires that have rough and viable areas in and around them are an important source of plaque accumulation that puts patients at risk for periodontal disease [19]. *S. mutans* is able to adhere to types of orthodontic brackets and form biofilms [20]. The model of the bracket and the reactions of the components of the method, including brackets, photosensitizers, and light, may affect the growth of bacteria and the effectiveness of the method [19]. However, a previous study has found no significant difference in *S. mutans* adhesion between different types of orthodontic brackets, including stainless steel, metal, composite, ceramic, and plastic brackets [21].

According to the findings of this study, the null hypothesis can be rejected. This study showed that curcumin + BDL with an output power of 500 mW was the most effective group, being as effective as CXH. CHX is a popular antimicrobial agent in dentistry that effectively eliminates microbes. It is the gold standard for the destruction of microbial biofilms [22]. However, it has some disadvantages, including taste impairment and discoloration of teeth and mucous membranes [23]. All groups receiving curcumin as a photosensitizer were more effective than groups receiving riboflavin, except for the curcumin alone, which showed no significant difference from riboflavin + BDL with an output power of 500 mW. The riboflavin alone and riboflavin + BDL with an output power of 200 mW groups showed no significant difference from the control group. However, neither group was able to completely kill all microorganisms.

One of the main reasons for the conflicting results within the groups is due to the properties of the photosensitizers. In this experiment, the antibacterial activity of aPDT with curcumin or riboflavin photosensitizers on *S. mutans* biofilms was evaluated by calculating CFU/mL. Araujo and coworkers studied the efficacy of curcumin in combination with blue LED on salivary bacteria of 13 adults and found a considerable decrease in bacterial viability [24]. Paschoal et al. also showed that the group receiving curcumin plus blue LED had a 70% greater reduction in bacterial counts than the control group under certain experimental conditions [25]. Our findings in the present study are consistent with those of previous studies. Kamran et al. [9] clearly demonstrated that riboflavin at a total concentration of 0.5% effectively reduced the viability of *Streptococcus sanguinis* and *S. mutans* biofilms around fixed orthodontic brackets using blue LED light with a power density of 2000 mW/cm^2, a wavelength range of 450 $\pm$ 65, and an energy density of 95 J/cm^2. The biofilms on the orthodontic brackets of the group receiving riboflavin alone were mostly viable. They showed that blue LED with riboflavin is as effective as CHX, but we did not observe such results. An important difference between Kamran et al. and our study is that they used metal brackets, whereas we used stainless steel brackets, which may have affected the results. Another study investigated the efficacy of riboflavin-mediated aPDT using BDL with different output powers (200–500 mW). They showed that irradiation of bacterial cultures with a BDL in the presence of riboflavin as a photosensitizer resulted in a dose-dependent decrease in *Enterococcus faecalis* viability [26]. The results of our study showed that the antibacterial potential of aPDT method increased with increasing strength (BDL), with curcumin + BDL with an output power of 500 mW showing the highest efficacy among the study groups. In agreement with our study, this study showed a greater reduction in *E. faecalis* at higher power of BDL. However, BDL alone at 200 mW to 500 mW showed no significant difference between the different output powers [26]. Other studies have shown that BDL or blue LED with riboflavin as a photosensitizer reduces colony numbers of *S. mutans*, *E. faecalis*, *Staphylococcus aureus*, and *Escherichia coli* to varying degrees [9,26–29]. The differences in the results on the efficacy of aPDT with riboflavin may be due to different bacterial strains, the site of the procedure, the concentration of the photosensitizer substance, the duration of the light source, and different laser parameters. However, Etemadi et al. have shown in a systematic review that there is no study reporting that riboflavin-mediated aPDT cannot reduce the number of bacterial colonies [30]. While we observed a large difference between riboflavin-containing groups and CHX, Morelato et al. recently showed that aPDT at 445 nm (Q power = 100 mW, 100 Hz, 124.34 W/cm^2, 1.24 J/cm^2) with a 0.1% riboflavin dye is as effective as 0.2% CHX. This contradiction may be explained by the microbial species used (*S. aureus* and *Candida albicans*) or the procedure [29].

A comparison of twelve groups and the different BDL irradiation powers were strengths of our study. Undoubtedly, this study has some limitations. The evaluation of antimicrobial capacity at a single time point is not sufficient to justify the potential antibacterial efficacy. Moreover, the present results could only be studied on a monomicrobial biofilm. Future studies should test the effect of the combination of BDL and curcumin/riboflavin on multiple microorganisms and determine the significance of curcumin/riboflavin concentration and BDL parameters for each strain tested. In addition, this study was designed as an in vitro study, which limits the transferability of the results to in vivo conditions and clinical setting. In vitro studies cannot determine environmental factors such as restricted accessibility, plaque formation, salivary flow, immune system influence, etc. Clinical studies are needed to obtain more reliable results. Another limitation of the study could be the relatively small sample size. Our results encourage further investigation of the in vitro and in vivo effects of novel disinfection therapies using curcumin/riboflavin and BDL. In addition, the shear bond strength of our method should be investigated in future studies.

5. Conclusions

This study showed that curcumin-mediated aPDT was more effective than riboflavin-mediated aPDT. The higher output powers were more effective in the reduction in *S. mutans* CFU/mL. The microbial decontamination potential of curcumin/riboflavin with BDL makes it possible to expand the scope of this device.

Author Contributions: Conceptualization, S.A., S.B., A.S. and N.C.; methodology, S.A. and N.C.; writing—original draft preparation, E.P. and T.G.; writing—review and editing, S.A. All authors have read and agreed to the published version of the manuscript.

Funding: This research received no external funding.

Institutional Review Board Statement: Not applicable.

Informed Consent Statement: Not applicable.

Data Availability Statement: Not applicable.

Conflicts of Interest: The authors declare no conflict of interest.

References

1. Wishney, M. Potential risks of orthodontic therapy: A critical review and conceptual framework. *Aust. Dent. J.* **2017**, *62*, 86–96. [CrossRef] [PubMed]
2. Ahirwar, S.S.; Gupta, M.; Snehi, S.K. Dental caries and lactobacillus: Role and ecology in the oral cavity. *Int. J. Pharm. Sci. Res.* **2019**, *11*, 4818–4829.
3. Qin, X.; Zi, H.; Zeng, X. Changes in the global burden of untreated dental caries from 1990 to 2019: A systematic analysis for the Global Burden of Disease study. *Heliyon* **2022**, *8*, e10714. [CrossRef] [PubMed]
4. Abrar, E.; Naseem, M.; Baig, Q.A.; Vohra, F.; Maawadh, A.M.; Almohareb, T.; AlRifaiy, M.Q.; Abduljabbar, T. Antimicrobial efficacy of silver diamine fluoride in comparison to photodynamic therapy and chlorhexidine on canal disinfection and bond strength to radicular dentin. *Photodiagn. Photodyn. Ther.* **2020**, *32*, 102066. [CrossRef] [PubMed]
5. Ahmadi, H.; Ebrahimi, A.; Ahmadi, F. Antibiotic therapy in dentistry. *Int. J. Dent.* **2021**, *2021*, 6667624. [CrossRef]
6. Cieplik, F.; Deng, D.; Crielaard, W.; Buchalla, W.; Hellwig, E.; Al-Ahmad, A.; Maisch, T. Antimicrobial photodynamic therapy—What we know and what we don't. *Crit. Rev. Microbiol.* **2018**, *44*, 571–589. [CrossRef]
7. Moro, G.G.; Massat, N.C.; Grandizoli, D.R.P.; Junior, A.E.; Degasperi, G.R.; Fontana, C.E.; Pinheiro, S.L. Effect of cetrimide 2% with and without photodynamic therapy to reduce *Streptococcus mutans* burden in dentinal carious lesions. *Lasers Med. Sci.* **2021**, *36*, 1935–1940. [CrossRef]
8. Anand, P.; Kunnumakkara, A.B.; Newman, R.A.; Aggarwal, B.B. Bioavailability of curcumin: Problems and promises. *Mol. Pharm.* **2007**, *4*, 807–818. [CrossRef]
9. Kamran, M.A.; Qasim, M.; Udeabor, S.E.; Hameed, M.S.; Mannakandath, M.L.; Alshahrani, I. Impact of riboflavin mediated photodynamic disinfection around fixed orthodontic system infected with oral bacteria. *Photodiagn. Photodyn. Ther.* **2021**, *34*, 102232. [CrossRef]
10. Comeau, P.; Burgess, J.; Qomi, N.R.; Lee, A.; Manso, A. The antimicrobial, physical, and chemical properties of a riboflavin-loaded dental resin intended for antimicrobial photodynamic therapy. *Photodiagn. Photodyn. Ther.* **2022**, *40*, 103124. [CrossRef]
11. Fawzy, A.S.; Nitisusanta, L.I.; Iqbal, K.; Daood, U.; Neo, J. Riboflavin as a dentin crosslinking agent: Ultraviolet A versus blue light. *Dent. Mater.* **2012**, *28*, 1284–1291. [CrossRef] [PubMed]
12. Subhan, M.A.; Alam, K.; Rahaman, M.S.; Rahman, M.A.; Awal, R. Synthesis and characterization of metal complexes containing curcumin (C21H20O6) and study of their anti-microbial activities and DNA-binding properties. *J. Sci. Res.* **2013**, *6*, 97–109. [CrossRef]
13. Priyadarsini, K.I. The chemistry of curcumin: From extraction to therapeutic agent. *Molecules* **2014**, *19*, 20091–20112. [CrossRef]
14. Leite, D.P.V.; Paolillo, F.R.; Parmesano, T.N.; Fontana, C.R.; Bagnato, V.S. Effects of photodynamic therapy with blue light and curcumin as mouth rinse for oral disinfection: A randomized controlled trial. *Photomed. Laser Surg.* **2014**, *32*, 627–632. [CrossRef]
15. Azizi, A.; Shohrati, P.; Goudarzi, M.; Lawaf, S.; Rahimi, A. Comparison of the effect of photodynamic therapy with curcumin and methylene Blue on streptococcus mutans bacterial colonies. *Photodiagn. Photodyn. Ther.* **2019**, *27*, 203–209. [CrossRef] [PubMed]
16. Lee, H.J.; Kang, S.M.; Jeong, S.H.; Chung, K.H.; Kim, B.I. Antibacterial photodynamic therapy with curcumin and Curcuma xanthorrhiza extract against Streptococcus mutans. *Photodiagn. Photodyn. Ther.* **2017**, *20*, 116–119. [CrossRef] [PubMed]
17. Tonon, C.C.; Paschoal, M.A.; Correia, M.; Spolidório, D.M.; Bagnato, V.S.; Giusti, J.S.; Santos-Pinto, L. Comparative effects of photodynamic therapy mediated by curcumin on standard and clinical isolate of *Streptococcus mutans*. *J. Contemp. Dent. Pract.* **2015**, *16*, 1–6. [PubMed]
18. Miles, A.A.; Misra, S.S.; Irwin, J.O. The estimation of the bactericidal power of the blood. *J. Hyg.* **2009**, *38*, 732–749. [CrossRef]
19. Antoun, J.S.; Mei, L.; Gibbs, K.; Farella, M. Effect of orthodontic treatment on the periodontal tissues. *Periodontol. 2000* **2017**, *74*, 140–157. [CrossRef]
20. Papaioannou, W.; Panagopoulos, A.; Koletsi-Kounari, H.; Kontou, E.; Makou, M. Adhesion of Porphyromonas gingivalis and biofilm formation on different types of orthodontic brackets. *Int. J. Dent.* **2012**, *2012*, 471380. [CrossRef]
21. Papaioannou, W.; Gizani, S.; Nassika, M.; Kontou, E.; Nakou, M. Adhesion of Streptococcus mutans to Different Types of Brackets. *Angle Orthod.* **2007**, *77*, 1090–1095. [CrossRef] [PubMed]
22. Balagopal, S.; Arjunkumar, R. Chlorhexidine: The Gold Standard Antiplaque Agent. *J. Pharm. Sci.* **2013**, *5*, 270.
23. Polizzi, E.; TetÃ, G.; Bova, F.; Pantaleo, G.; Gastaldi, G.; CapparÃ, P.; Gherlone, E. Antibacterial properties and side effects of chlorhexidine-based mouthwashes. A prospective, randomized clinical study. *J. Osseointegr.* **2019**, *12*, 2–7.
24. Araújo, N.C.; Fontana, C.R.; Gerbi, M.E.; Bagnato, V.S. Overall-mouth disinfection by photodynamic therapy using curcumin. *Photomed. Laser Surg.* **2012**, *30*, 96–101. [CrossRef]
25. Paschoal, M.A.; Tonon, C.C.; Spolidório, D.M.; Bagnato, V.S.; Giusti, J.S.; Santos-Pinto, L. Photodynamic potential of curcumin and blue LED against *Streptococcus mutans* in a planktonic culture. *Photodiagn. Photodyn. Ther.* **2013**, *10*, 313–319. [CrossRef] [PubMed]
26. Afrasiabi, S.; Chiniforush, N. Antibacterial potential of riboflavin mediated blue diode laser photodynamic inactivation against Enterococcus faecalis: A laboratory investigation. *Photodiagn. Photodyn. Ther.* **2023**, *41*, 103291. [CrossRef] [PubMed]
27. Nielsen, H.K.; Garcia, J.; Væth, M.; Schlafer, S. Comparison of riboflavin and toluidine Blue O as photosensitizers for photoactivated disinfection on endodontic and periodontal pathogens in vitro. *PLoS ONE* **2015**, *10*, e0140720. [CrossRef]

28. Banerjee, S.; Ghosh, D.; Vishakha, K.; Das, S.; Mondal, S.; Ganguli, A. Photodynamic antimicrobial chemotherapy (PACT) using riboflavin inhibits the mono and dual species biofilm produced by antibiotic resistant *Staphylococcus aureus* and *Escherichia coli*. *Photodiagn. Photodyn. Ther.* **2020**, *32*, 102002. [CrossRef]
29. Morelato, L.; Budimir, A.; Smojver, I.; Katalinić, I.; Vuletić, M.; Ajanović, M.; Gabrić, D. A novel technique for disinfection treatment of contaminated dental implant surface using 0.1% riboflavin and 445 nm diode laser-an in vitro study. *Bioengineering* **2022**, *9*, 308. [CrossRef]
30. Etemadi, A.; Hamidain, M.; Parker, S.; Chiniforush, N. Blue light photodynamic therapy with curcumin and riboflavin in the management of periodontitis: A Systematic Review. *J. Lasers Med. Sci.* **2021**, *12*, e15. [CrossRef]

biomedicines

MDPI

Article

Different Effects of Phototherapy for Rat Glioma during Sleep and Wakefulness

Alexander Shirokov [1,2,*], Inna Blokhina [2], Ivan Fedosov [3], Egor Ilyukov [3], Andrey Terskov [2], Dmitry Myagkov [3], Dmitry Tuktarov [3], Maria Tzoy [3], Viktoria Adushkina [2], Daria Zlatogosrkaya [2], Arina Evsyukova [2], Valeria Telnova [2], Alexander Dubrovsky [3], Alexander Dmitrenko [2], Maria Manzhaeva [2], Valeria Krupnova [2], Matvey Tuzhilkin [2], Inna Elezarova [2], Nikita Navolokin [2,4], Elena Saranceva [2], Tatyana Iskra [2], Ekaterina Lykova [2] and Oxana Semyachkina-Glushkovskaya [2,5,*]

1. Institute of Biochemistry and Physiology of Plants and Microorganisms, Saratov Scientific Centre of the Russian Academy of Sciences, Prospekt Entuziastov 13, 410049 Saratov, Russia
2. Department of Biology, Saratov State University, Astrakhanskaya Str. 83, 410012 Saratov, Russia; inna-474@yandex.ru (I.B.); terskow.andrey@gmail.com (A.T.); adushkina.info@mail.ru (V.A.); eloveda@mail.ru (D.Z.); arina-evsyukova@mail.ru (A.E.); ler.vinnick2012@yandex.ru (V.T.); admitrenko2001@mail.ru (A.D.); mariamang1412@gmail.com (M.M.); krupnova_0110@mail.ru (V.K.); tuzhilkinma@yandex.ru (M.T.); inna.elizarowa7@yandex.ru (I.E.); nik-navolokin@yandex.ru (N.N.); sophora68@mail.ru (E.S.); tata-isk@yandex.ru (T.I.); eckaterina_lykova@mail.ru (E.L.)
3. Physics Department, Saratov State University, Astrakhanskaya Str. 83, 410012 Saratov, Russia; fedosov_optics@mail.ru (I.F.); egor.re01@mail.ru (E.I.); dmyagk0v@yandex.ru (D.M.); ivanov.ivao@yandex.ru (D.T.); dethaos@bk.ru (M.T.); paskalkamal@mail.ru (A.D.)
4. Department of Pathological Anatomy, Saratov Medical State University, Bolshaya Kazachaya Str. 112, 410012 Saratov, Russia
5. Physics Department, Humboldt University, Newtonstrasse 15, 12489 Berlin, Germany

* Correspondence: shirokov_a@ibppm.ru (A.S.); glushkovskaya@mail.ru (O.S.-G.)

Abstract: There is an association between sleep quality and glioma-specific outcomes, including survival. The critical role of sleep in survival among subjects with glioma may be due to sleep-induced activation of brain drainage (BD), that is dramatically suppressed in subjects with glioma. Emerging evidence demonstrates that photobiomodulation (PBM) is an effective technology for both the stimulation of BD and as an add-on therapy for glioma. Emerging evidence suggests that PBM during sleep stimulates BD more strongly than when awake. In this study on male *Wistar* rats, we clearly demonstrate that the PBM course during sleep vs. when awake more effectively suppresses glioma growth and increases survival compared with the control. The study of the mechanisms of this phenomenon revealed stronger effects of the PBM course in sleeping vs. awake rats on the stimulation of BD and an immune response against glioma, including an increase in the number of CD8+ in glioma cells, activation of apoptosis, and blockage of the proliferation of glioma cells. Our new technology for sleep-phototherapy opens a new strategy to improve the quality of medical care for patients with brain cancer, using promising smart-sleep and non-invasive approaches of glioma treatment.

Keywords: glioma; photobiomodulation; brain drainage; CD8+ cells; immune response

1. Introduction

Sleep deficit is a common symptom of glioma, with growing evidence suggesting an association between sleep disturbance and poor physical and psychological outcomes, including survival [1–6]. One study using the General Sleep Disturbance Scale reported insomnia (disruption in sleep quantity, pattern, or architecture) in 100% of patients with glioma [7]. Other studies estimated the prevalence of sleep injuries in patients with glioma to be between 37.0 and 81.8% [8–10]. It is unknown whether sleep deficiency can be reliably correlated with glioma growth, but there are growing numbers of case reports to suggest that a connection is possible [6,11–14]. Indeed, Sadighi et al. found that specific sleep disturbances are contingent on the tumor's location [12]. Glioma location is also an important factor in choosing a treatment plan. The sleep centers can be affected by glioma directly or by neuroinflammation and peritumoral edema from a growing tumor [11,13]. Subramanian et al. proposed that specific sleep disturbance induced by glioma growth could be a systemic marker of the initial location of glioma for surgical resection and be used as a potential tool to track recurrence [13]. The correlation between sleep disturbance, location of brain tumors, and survival is discussed in several excellent reviews [5,6,11,15]. However, its merits in the treatment of glioma and its recurrence have yet to be explored.

A decrease in resistance to glioma progression during the development of sleep deficiency may be associated with the suppression of brain drainage (BD), the activity of which is strictly dependent on sleep [16–22]. The reduced BD in subjects with glioma could be a reason for the accumulation of excess fluid in the skull, leading to a dramatic increase in the intracranial pressure [23]. Indeed, peritumoral edema promotes the accumulation of extensive brain fluids that is associated with high mortality in patients with glioma [23]. The reasons for the development of cerebral edema in glioma remain unknown. There is emerging evidence that reduced BD can play a crucial role in glioma progression [16–18]. Indeed, Ma et al. revealed that outflow of the cerebrospinal fluid (CSF) is reduced in glioma due to a blockage of circulation of the cerebral spinal fluid (CSF) [16]. The extensive accumulation of brain fluids aggravates the brain tumor microenvironment and attenuates intracranial drug delivery efficacy. The development of methods for the compensation of CSF outflow and restoring normal CSF circulation are worthy of clinical attention.

Recently, transcranial photobiomodulation (PBM) has been shown to effectively stimulate BD and the lymphatic removal of toxins from the brain [21,22,24–28]. There is evidence that the stimulating effects of PBM on BD are achieved by PBM-mediated regulation of the contraction and relaxation phases of the different lymphatic vessels (LVs) that we analyzed and discussed in our previous studies [26–28]. Indeed, using in vivo monitoring of cervical LVs carrying lymph from the brain to the neck lymph nodes or in the mesenteric lymphangion, we clearly showed PBM-induced stimulation of lymphatic contractility [27,28]. Furthermore, in in vitro experiments, we demonstrated that PBM induces a transient increase in the NO level in the culture of lymphatic endothelial cells obtained from the mesenteric lymphatics [28]. In ex vivo experiments, we found the PBM effects on the basal MLVs are suppressed by the blockage of NO generation [28]. Thus, PBM has both effects on the different LVs: stimulation of their constriction and an increase in NO production. The lymphatics produce NO during contraction as flow shear activates the endothelial cells [29–31]. The elevation of NO then contributes to the subsequent relaxation of different types of peripheral LVs. Traditionally, it was believed that NO suppressed lymphatic constriction. However, the NO-mediated regulation of lymphatics is not as simple as was first assumed. So, the basal NO in the mesenteric lymphatics increases with the frequency of contraction induced by systemic administration of saline [29]. In effect, elevated lymphatic pumping increases the NO production due to increased flow shear forces. In this case, NO provides a generalized inhibition of pumping during periods of high lymph flow. Thus, NO supports the relaxation of lymphatics after its constriction. The NO effects are different along LVs. The high NO production during lymphatic constriction is observed predominantly in the valves [29]. Indeed, the mesenteric lymphatic valves contain a 30–50% higher NO concentration than tubular regions during contraction due both to there being many endothelial cells and an increased expression of endothelial nitric oxide synthase. The NO generation in the lymphatic valves limits the pumped flow of the total lymphatics by lowering the frequency and stroke volume of individual contractions. Thus, despite the fact that all lymphatic endothelial cells are capable of generating NO in flow shear events during contractions [29], this does not mean that the role of NO is uniformly important for all sections of LVs during the contraction and relaxation process. From a physiological point of view, NO generated in the lymphatic valves during constriction diffuses into the flowing lymph, survives to downstream tubular sites, and contributes to tubular relaxation that is important for the peristaltic character of pumping [30,31].

Strikingly, PBM also suppresses glioma growth [24]. There is pioneering data demonstrating that PBM increases the resistance of rats to glioma growth and the survival of animals due to PBM-induced activation of apoptosis of glioma cells and a reduction in intracranial pressure through stimulation of BD [24]. Therefore, PBM may be a promising therapeutic approach in non-invasive treatment of glioma via photo-activation of the MLV functions and BD.

Taking into account all of the above, we have focused on solving the problem of the improvement of BD in Wistar rats with glioma using an innovative approach of PBM for MLVs under EEG control of sleep and wakefulness. The steroids are standard therapy for peritumoral edema surrounding gliomas [32]. However, steroid therapy has limited efficacy and significant side effects [33]. New therapeutic strategies targeting BD and the MLV functions appear to be promising [16–18]. The re-discovery of MLVs prompted a re-evaluation of the mechanisms responsible for regulation of BD and the formation of cerebral edema in gliomas [16–18,34]. In our series of experiments, for the first time, we clearly demonstrated that PBM stimulates BD in both healthy rodents and rats with glioma [24–28]. By studying the mechanisms of the therapeutic effects of PBM in the in vivo, ex vivo, and in vitro experiments, we discovered that PBM increases the contractility of LVs and increases their drainage properties [26–28]. Quite recently, in our study on mice with Alzheimer's disease, we discovered that the effects of PBM on BD are stronger in sleeping vs. awake animals [21,22]. This phenomenon can be explained by the fact that during deep sleep, BD is activated due to an increase in the size of the perivascular spaces, which promotes the removal of toxins and metabolites dissolved in CSF [19]. Since in our previous work using PBM in awake rats we proved the suppression of glioma growth, we hypothesized that sleep could enhance the therapeutic effects of PBM due to better activation of BD. To test this hypothesis, in this study we compared the PBM effects during sleep and wakefulness on the glioma growth, including the apoptosis and proliferation of tumor cells, as well as on survival rate and BD, in Wistar male rats.

2. Materials and Methods

2.1. Subjects

Pathogen-free male *Wistar* rats (200–250 g, 2 months old) were used in all experiments and were obtained from the National Laboratory Animal Resource Centre (Shemyakin-Ovchinnikov Institute of Bioorganic Chemistry, RAS, Pushchino, Russia). The animals were housed under standard laboratory conditions with access to food and water ad libitum. All experimental procedures were performed in accordance with the "Guide for the Care and Use of Laboratory Animals", Directive 2010/63/EU on the Protection of Animals Used for Scientific Purposes, and the guidelines from the Ministry of Science and High Education of the Russian Federation (№ 742 from 13 November 1984), which have been approved by the Bioethics Commission of the Saratov State University (Protocol No. 8, 18 April 2023). The mice were housed at 25 ± 2 °C, 55% humidity, and subject to a 12:12 h light–dark cycle (light: 08:00 a.m.–08:00 p.m.). The mice adapted to the experimental conditions during one week before the beginning of the experiments to ensure acclimation to the housing room of the animal facility. The experiments were performed in the following groups: (1) control (healthy rats) without the PBM course; (2) sham rats without the PBM course; (3) sham rats receiving the PBM course under EEG control of wakefulness; (4) sham rats receiving the PBM course under EEG control of deep sleep; (5) rats with glioma without the PBM course; (6) rats with glioma receiving the PBM course under EEG control of wakefulness; (7) rats with glioma receiving the PBM course under EEG control of deep sleep; n = 10 in each group in all sessions of the experiments and n = 20 in the study of survival rate.

2.2. Model of Rat Glioma

The C6 rat glioma cell line was obtained from the Russian Cell Culture Collection of Vertebrates, Institute of Cytology, Russian Academy of Sciences (St. Petersburg, Russia). A transfected C6—TurboRFP cell line was used for the study of the growth of fluorescent GBM [24]. The C6 cells were cultured in a Dulbecco's Modified Eagle Medium (DMEM) growth medium (Paneco, Moscow, Russia) containing 2.5% embryonic veal serum (Biosera, Cholet, France), 4 mM glutamine (Paneco, Moscow, Russia), penicillin (50 IU/mL), and streptomycin (50 mg/mL) (Paneco, Moscow, Russia). Rat C6 glioma cells were transfected with TurboRFP-C DNA plasmids using the method of liposomal transfection followed by selection using geneticin (G418 antibiotic, neomycin analogue). The resulting cell line, C6—TurboRFP, has stable cultural and morphological characteristics.

The glioma cells (5×10^5 cells per rat) were injected at coordinates AP—1 mm, ML—1 mm, DV—4 mm, with a Hamilton microsyringe in a volume of 15 μL at a rate of 1 μL/min. Physiological saline (15 μL, Sigma-Aldrich, St. Louis, MO, USA) was injected in the same region of the brain in the sham groups. Thereafter, the burr hole was sealed with sterile bone wax and tissue glue and the wound was sutured closed with 3–0 absorbable suture material. After the implantation of glioma cells, the wound was closed and treated with 2% brilliant green solution. The rats were removed from the stereotaxic head holder, given 0.01 mg/kg buprenorphine, s.c., and 50 K bicillin, i.m., returned to a temperature-controlled recovery cage, and moved back to the animal facility after recovery. Glioma growth and tumor volume were monitored using magnetic resonance imagining (MRI) 28 days after the tumor cell implantation using a Clin scan 7T tomograph (Bruker, Mannheim, Germany). The growth of fluorescent glioma C6—TurboRFP in the sham group and in rats that received the PBM course during wakefulness or sleep was assessed using confocal microscopy using a Nikon A1R MP confocal laser scanning microscope (Nikon Corp., Tokyo, Japan). For this purpose, 4 weeks after implantation of glioma cells, the brains of rats from the tested groups, including the control without PBM and the experimental groups that received PBM while asleep or awake, were removed and a whole fresh set of brains was scanned.

2.3. PBM under EEG Control

In this study, we used our adapted and previously published protocol for PBM under EEG control [35]. A two-channel cortical EEG was recorded. The rats were implanted with two silver electrodes (tip diameter: 2–3 μm) located at a depth of 150 μm at coordinates ML: 3.0 mm and AP: 3.0 mm from the bregma on either side of the midline under inhalation anesthesia with 1% isoflurane (Sigma-Aldrich, St. Louis, MO, USA), at a rate of 1 L/min (N_2O/O_2—70/30 ratio). The head plate was mounted and small burr holes were drilled. Afterward, EEG wire leads were inserted into the burr holes on one side of the midline between the skull and the underlying dura. The EEG leads were secured with dental acrylic (Zhermack SpA, Badia Polesine, Venice, Italy). Ibuprofen (Bhavishya Pharmaceuticals Pvt. Ltd., Hyderabad, Telangana, India; 15 mg/kg), for the relief of postoperative pain, was provided in the rats' water supply for 2 to 3 days prior to surgery and for 3 days post-surgery. The rats were allowed 10 days to recover from surgery prior to beginning the experiment.

EEG recording was performed with a tethered EEG system using an ADS1293 (Texas instruments, Dallas, TX, USA) low-power, 3-channel, 24-bit analog front-end for biopotential measurements. Initialization and data transfer were performed with an Atmega328 (Atmel, San Jose, CA, USA) microcontroller via an SPI interface. The same microcontroller was used to detect non-rapid eye movement (NREM) sleep in real time. The instrument was powered with an 18650 Li-ion battery. An ESP-01 (Espressif Systems, PRC) module was used for Wi-Fi communication between the instrument and standalone PC. The instrument was placed on top of the home cage and connected to the mouse with a 0.3 m long flexible 4-wire cable attached to a head-mounted miniature connector soldered to 4 screw electrodes. The connector allowed for the EEG instrument to be easily plugged into the mouse without anesthesia (Figure 1a). Note that anesthesia strongly affects the brain's functions and BD, which makes it necessary to avoid the use of anesthesia in our experiments [36].

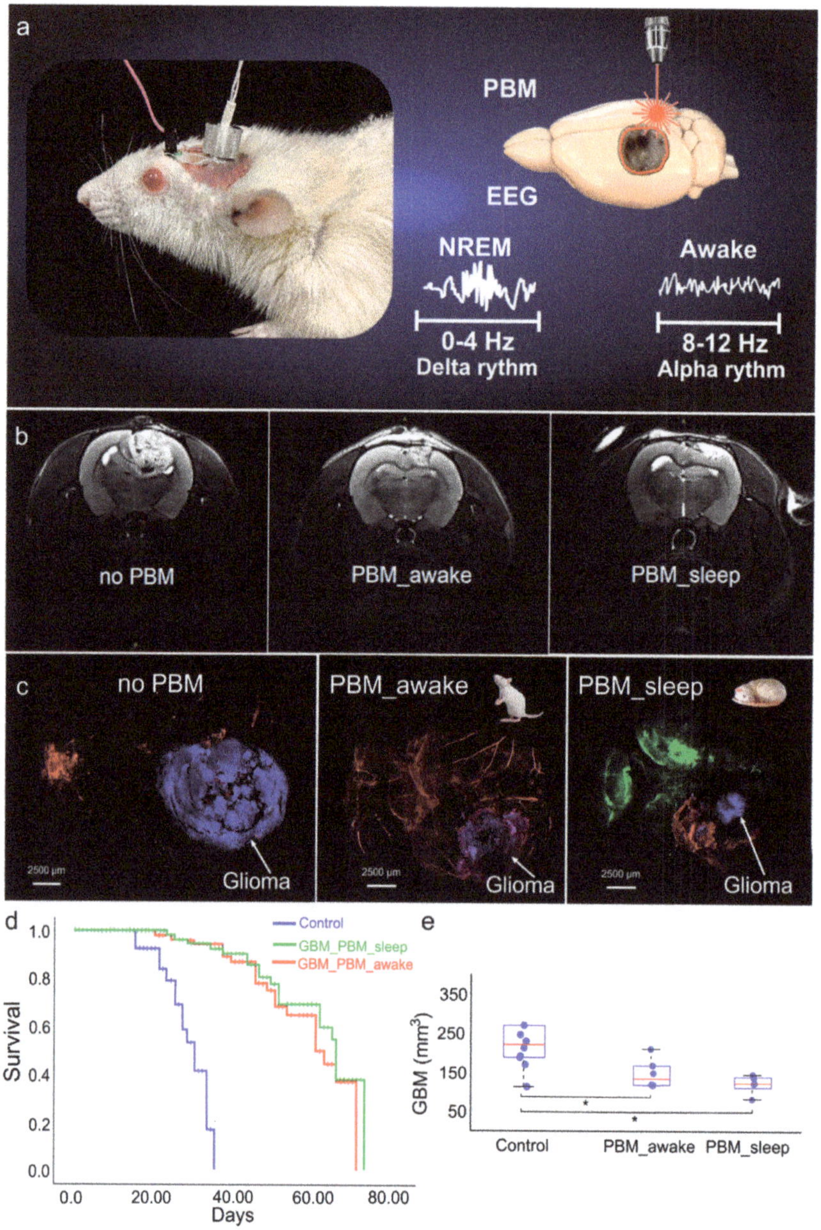

Figure 1. The effects of PBM on survival and progression of rats' glioma: (**a**) schematic illustration of PBM under EEG control during sleep or awake states; (**b**) representative images of ex vivo confocal analysis of glioma size in rats without and after the PBM course during wakefulness or sleep, respectively; (**c**) representative MRI images of rat glioma 4-week growth without and after the PBM

course during wakefulness or sleep; (**d**) Kaplan–Meier overall survival plots between the tested groups; the survival differences between rats with glioma, without, and after the course of PBM during sleep or wakefulness was significant (X2 test, p = 0.00000000000129 between the no PBM and the PBM_sleep groups and p = 0.0000000000024 between the no PBM and the PBM_awake groups; n = 20 in each group); (**e**) MRI analysis of glioma size in rats without and after the course of PBM during sleep or wakefulness; n = 10 in each group; *—$p < 0.05$ between groups, the ANOVA test with post hoc Duncan test.

A 1050 nm LED with an output power of 50 mW in 2853 SMD housing was used for PBM. The LED driver was controlled with the pulse-width modulation (PWM) output of the EEG instrument microcontroller. The LED was connected to the instrument with a 0.3 m long 2-wire flexible cable and mounted into a miniature 3D printed frame with a pair of cylindrical magnets 3 mm in diameter and 3 mm in height each. That allowed for easy attachment of the LED frame to an M3 steel washer glued at the surface of the mouse's skull. The LED operated in PWM mode at a 1 kHz modulation frequency. The washer hole, of 3.5 mm in diameter, acted as an aperture limiting the PBM area to 0.1 cm^2 at the skull's surface.

The LED output power of optical radiation was 50 mW and the area limited with a washer was 0.1 cm^2. Thus, in continuous operation mode the power density was 500 mW/cm^2, complying with ANSI Z136 for maximal permissible exposure (MPE) of skin to 1050 nm radiation. When operating in the pulse-width modulation (PWM) regime at 1 kHz with a duty cycle less than 10%, the instant and averaged power densities were both at least 10 times less than the MPE. A duty cycle of 2% corresponds to 10 mW/cm^2. For a 17 min (1020 s) long procedure at 2% PWM duty, the LED delivers a 10 J/cm^2 dose over the irradiated skull's surface. The instrument was controlled with software developed using LabVIEW (NXG 5.1, National instruments, Dallas, TX, USA).

The software enables EEG recording, monitoring of the instrument's operation, and remote configuration of the NREM sleep detection and PBM procedure. The EEG signal was digitized at 2.4 kSa/s, and filtered with a fifth-order digital Butterworth bandpass filter with a lower cutoff frequency of 0.5 Hz and an upper one of 250 Hz. A detailed description of the method is given in our article [35]. When 20% of the 30 s epochs (6 of 30 consecutive 1 s records) were scored as NREM [37], this was defined as the NREM sleep stage of the animal, and the PBM LED was automatically turned on to run for 17 min at the given PBM. Once configured, the instrument was operated autonomously, logging its operations to be monitored via a Wi-Fi connection. Wakefulness, NREM, and rapid eye movement (REM) sleep were defined as described in our previous studies, where we demonstrate the EEG patterns and spectrum characteristics of wakefulness, NREM, and REM sleep in rodents [21,22].

2.4. Optical Monitoring of Brain's Drainage System

To study the PBM effects on BD, we analyzed spreading of fluorescein isothiocyanate (FITC)-dextran (FITCD, Sigma, St. Louis, MO, USA) in the dorsal and ventral parts of the whole brain after a single PBM application using ex vivo confocal microscopy. An amount of 7 μL of FITCD was injected into the right lateral ventricle (AP—1.0 mm; ML—1.4 mm; DV—3.5 mm) at a rate of 0.1 μL/min using a microinjector (Stoelting, St. Luis, USA) with a Hamilton syringe with a 29-G needle (Hamilton Bonaduz AG, Switzerland). The implantation of a chronical polyethylene catheter (PE-10, 0.28 mm ID × 0.61 mm OD, Scientific Commodities Inc., Lake Havasu City, AZ, USA) into the right lateral ventricle was preformed according to the protocol reported by Devos et al. [38].

The head plate with LED was fixed in the region of the parietal and interparietal bones using dental acrylic (Zhermack SpA, Badia Polesine, Italia) under inhalation anesthesia with 1% isoflurane (Sigma-Aldrich, St. Louis, MO, USA), at a rate of 1 L/min (N_2O/O_2—70/30 ratio). The LED was fixed to the head plate with two screws and was placed on the parietal region, where the largest number of MLVs is located [39] (Figure 1a). The LED was performed for 61 min using the following sequence, 17 min—LED, 5 min—pause,

17 min—LED, 5 min—pause, 17 min—LED, as we have determined by a random selection in our previous studies [21,22,24,35]. The irradiance at the skull's surface did not exceed 0.5 W/cm^2. The dose for PBM during a single PBM treatment was 30 J/cm^2 and for the 14 days of the PBM course it was 0.42 kJ/cm^2. Using this PBM dose, we did not find any changes in temperature on the brain surface after PBM, consistent with the results of our previous study [24].

A type A-K3 thermocouple (Ellab, Hillerød, Denmark) was used to measure the skull's temperature. The thermocouple was placed subcutaneously 2 mm lateral to the bregma in the irradiated zone. A burr hole was drilled under inhalation anesthesia (1% isoflurane at 1 L/min; N_2O/O_2—70:30). To measure the brain surface temperature under 1050 nm LED irradiation, the medial part of the left temporal muscle was detached from the skull bone, a small burr hole was drilled into the temporal bone, and a flexible thermocouple probe (IT-23, 0.23 mm diameter, Physitemp Instruments LLC, Clifton, NJ, USA) was introduced between the parietal bone and the brain into the epidural space. Brain surface temperature was measured before and during laser stimulation with a 5 min increment using a handheld thermometer (BAT-7001H, Physitemp Instruments LLC, Clifton, NJ, USA).

To study the PBM effects on BD during wakefulness and deep sleep, the intracerebroventricular injection of FITCD was performed at 08:00 a.m. under the EEG monitoring when awake and after 08:00 p.m. at the time of EEG monitoring of NREM for a 3 h observation period. The times of 8 a.m. and 8 p.m. for the intracerebroventricular injection of FITCD were chosen due to the light regime of the vivarium and to standardize the protocol to start the experiment at the time of the natural transition to sleep or wakefulness. To keep the same time for the distribution of FITCD in the awake and sleeping states, rats that did not show NREM during the 3 h observation period were not included in the studies. The ex vivo optical study of the FITCD distribution in the brain fluid system was performed 3 h after the intracerebroventricular injection of FITCD. Afterward, whole-brain imaging from the dorsal and ventral aspects as well as the deep cervical lymph nodes (dcLNs) was performed using confocal microscopy (Nikon Corp., Tokyo, Japan).

The 14-day PBM course during sleep or wakefulness was performed daily in rats with 4-week-old glioma. The PBM course under ECG control was performed during observation of NREM (synchronized activity with high amplitude, which is dominated by low-frequency delta waves (0–4 Hz) comprising >30% of EEG waveforms/epochs) or awake (low-amplitude and high-frequency dynamics >10%, 8–12 Hz).

The imaging was performed using a Nikon A1R MP upright confocal microscope (Nikon Corp., Tokyo, Japan) with CFI Plan Apochromat Lambda D 2X (Nikon Corp., Tokyo, Japan) installed in a focusing nosepiece. The brain samples were submerged in a buffer solution in a Petri dish placed on the microscope stage. The top surface of each sample was covered with a 25 mm × 25 mm × 0.17 mm glass cover slide. The brain image was captured as a stack of 5 stitched large images each over 6 × 6 fields of view; the dcLNs image was captured as a stack of 4 stitched large images each over 2 × 2 fields of view. The image resolution of the brain was 3636 × 3636 pixels at 6.11 μm/pixel and for dcLN 0.3218 μm/pixel. The Z step was 222 μm. The resulting image was obtained as a maximum-intensity Z projection of all 5 images of the stack. The method enables a high-resolution image to be obtained extended along a 1 mm depth of focus. The confocal images were captured in two channels: 488 nm excitation/500–550 emission was used to image the FITCD distribution; and 640 nm excitation/663–738 nm emission for Evans Blue dye (1%, Sigma-Aldrich, St. Louis, MO, USA) that was intravenously injected 30 min before the experiments for labeling of the cerebral blood vessels. The dcLNs were imaged identically but only a single field of view was captured for each stack. Image processing was performed using the Fiji open-source image-processing package [40]. The image-processing procedures were identical for each pair of images (control and laser-treated samples) for each channel to ensure an accurate comparison of the fluorescence intensity.

2.5. MRI of Rat Glioma

A tumor-volume assessment was performed in the 4th week after GBM implantation using a 7 Tesla Bruker BioSpec 70/30 USR dedicated research MRI scanner and the Paravision 6.0 data acquisition software (Bruker Biospin; Billerica, MA, USA). To obtain a good signal-to-noise ratio, a 72 mm small-bore linear RF volume coil with an actively decoupled brain surface coil (40 cm bore; 660 mT/m, rise time within 120 µs) was used for excitation and signal detection. Rats were anesthetized with 1% isoflurane at 1 L/min N_2O/O_2 (70:30). Temperature and respiration were monitored and maintained using a thermal air blower. Anatomical T2-weighted images were acquired with a fast spin echo sequence (rapid acquisition with relaxation enhancement; repetition time (TR)/echo time = 5000 ms/56 ms; field of view = 4 cm × 4 cm; slice thickness = 1 mm; slice gap (inter-slice distance) = 1.1 mm; number of slices = 12; matrix = 256 × 256; number of averaging = 3) as previously described [41]. T1-weighted imaging used the RARE technique, with 9.6 ms TE, 1000 ms TR, and a RARE factor of 2, thus 4 averages, requiring 4 min 16 s to assess tumor volumes; ROIs were drawn around regions of visible hyperenhancement in each of the slices on the T2-weighted and corresponding T1-weighted images using NIH ImageJ and calculated using the MATLAB software (version 2018b, MathWorks, Inc., Natick, MA, USA). Statistical analysis was performed using GraphPad Prizm v.6.0.

2.6. Immunohistochemistry (IHC)

Rats were euthanized with an intraperitoneal injection of a lethal dose of ketamine and xylazine and intracardially perfused with 0.1 M of PBS for 5 min. Afterward, their brains were removed and fixed in 10% buffered formalin with wiring material in alcohol, and poured into paraffin. The paraffin sections were stained with hematoxylin and eosin, and IHC studies were performed using the REVEAL Polyvalent HRP-DAB detection system. The monoclonal antibodies (Abcam, Cambridge, UK) Bcl (MAB8272), Ki67 (clone SP6, ab16667), and p53 (ab131442) were used at a dilution of 1:100 to the antibody. When staining with ICH markers, positive and negative controls were used to exclude false-negative and false-positive results, to create standardization of the staining conditions and increase the objectivity of the results. The percentage of positively expressing cells in 10 fields of view for each sample and the intensity of the immunohistochemical reactions were calculated. All studies were performed using a MicroVisor of medical transmitted light µVizo-103 (LOMO, St. Petersburg, Russia) with a magnification of 640 times.

For confocal imaging of the brain slices and the dcLNs we used the protocol for the IHC analysis with the markers for the transmembrane glycoprotein surface marker CD8+ T lymphocytes (CD8) for lymphatic vessel endothelial hyaluronan receptor 1 (LYVE1).

For the IHC analysis, brain and dcLN tissues were collected and free-floating sections were prepared. Tissue of the brain and dcLN were fixed for 48 h in a 4% saline-solution-buffered formalin, then sections of the brain and dcLN with a thickness of 40–50 microns were cut on a vibrotome (Leica, Wetzlar, Germany). The tissues of dcLN were previously poured into 2% agarose on saline solution. The sections were processed according to the standard immunohistochemical protocol with the corresponding primary and secondary antibodies. The sections of mouse brain and dcLN were imaged using a Leica SP8 confocal laser scanning microscope (Leica, Wetzlar, Germany). The antigen expression was evaluated on free-floating sections according to the standard method of simultaneous combined staining (Abcam protocols for free-floating sections).

The nonspecific activity was blocked by 2 h of incubation at room temperature with 10% BSA in a solution of 0.2% Triton X-100 in PBS. Solubilization of cell membranes was carried out during 1 h of incubation at room temperature in a solution of 1% Triton X-100 in PBS. The sections of the brain and of the dcLNs were, firstly, washed 3 times (5 min for each) with wash solution (0.2% Triton X-100 in PBS), secondly incubated in the blocking solution (a mixture of 0.2% Triton X-100 and 10% BSA in PBS) for 2 h, followed by incubation with rat anti-CD8+ antibody (1:500; ab 22378, Abcam, Cambridge, UK) and rabbit anti-Lyve-1 antibody (1:500; ab 218535, Abcam, Cambridge, UK) overnight at 4 °C in PBS containing

0.2% Triton X-100 and 0.5% normal goat serum. Next, samples were incubated at room temperature for 1 h and then washed 3 times, followed by incubation with goat anti-rat IgG (H+L) Alexa Fluor 488 (Invitrogen, Molecular Probes, Eugene, OR, USA), and goat anti-rabbit IgG (H+L) Alexa Fluor 555 (Invitrogen, Molecular Probes, Eugene, OR, USA) for visualizing the LVs in the dcLNs. In the final stage, the sections were transferred to the glass and 15 μL of mounting liquid (50% glycerin in PBS with DAPI at a concentration of 2 μg/mL) was applied to the section, then the preparation was covered with a cover glass and confocal microscopy was performed.

The sections of brains and dcLN were visualized using a confocal microscope (Nikon A1R MP, Nikon Instruments Inc., Melville, New York, USA) with a ×20 lens (0.75 NA), or a ×100 lens for immersion in oil (0.45 NA). DAPI, Alexa Fluor 488, and Alexa Fluor 555 were excited with excitation wavelengths of 405 nm, 488 nm, and 561 nm, respectively. Alexa Fluor 647 and Evans Blue were excited with the same excitation wavelength of 647 nm. Three-dimensional visualization data were collected by obtaining images in the x, y, and z planes. The images were obtained using the NIS-Elements software (v.5.21, Nikon Instruments Inc.) and analyzed using Fiji software (v.2.9.0, open-source image processing software) and Vaa3D (v. 1.1.4, open-source visualization and analysis software).

2.7. Statistical Analysis

All statistical analyses were performed using the Microsoft Office Excel and SPSS 17.0 for Windows software. The results were reported as a mean value ± standard error of the mean (SEM). The inter-group differences in all series experiments were evaluated using the ANOVA test with post hoc Duncan test. The significance levels were set at $p < 0.05$ for all analyses. No statistical methods were used to predetermine the sample size.

3. Results

3.1. The Effects of the PBM Course during Sleep or Awake State on Survival and Glioma Progression

In our previous study, we clearly showed that the therapeutic effects of PBM are higher in sleeping vs. awake animals with Alzheimer's disease [21,22]. Since in our recent investigation on awake rats we discovered that PBM might also be a promising tool for the suppression of glioma growth, here we compare the therapeutic effects of PBM during the sleep and awake states.

In the first step, we studied the effect of the PBM course on the survival rate and the tumor volume in sleeping and awake rats. The MRI data clearly show that the glioma volume was significantly decreased in both the PBM groups vs. the no PBM group (Figure 1b,e). However, the glioma volume reduction was 1.3-fold greater in rats receiving PBM during sleep than during wakefulness (Figure 1b,e).

Additionally, we performed confocal analysis of the glioma volume in the tested groups using our original model of fluorescent glioma (Figure 1c). The ex vivo confocal data also demonstrate that the effects of PBM on the suppression of glioma growth was significantly higher in the PBM_sleep group vs. the PBM_awake group.

The survival rate was evaluated using the Kaplan–Meier method. Figure 1d clearly demonstrates that the survival rate was significantly higher in both the PBM groups than in rats with glioma without PBM (X2 test, $p = 0.0000000000129$ between the no PBM and the PBM_sleep groups and $p = 0.00000000000024$ between the no PBM and the PBM_awake groups; $n = 20$ in each group). There were no differences in survival between rats receiving PBM during the sleep and awake states.

This series of experiments shows that although the courses of PBM in sleeping and awake rats increased their survival equally well, the decrease in glioma volume was more pronounced in rats receiving PBM in the sleep than in the awake state.

3.2. The Effects of the PBM Course during Sleep or Awake State on Apoptosis and Proliferation of Glioma Cells

The activation of the apoptosis of glioma cells and the suppression of their proliferation underlie the suppressive effects of PBM on the glioma growth, as we established in our recent studies in awake rats [24]. Hypothesizing that PBM during sleep could have stronger effects on the glioma progression than during the awake state, in the second step, we studied the cellular mechanisms of the therapeutic effects of the PBM course in sleeping and awake rats with glioma (Table 1, Figure 2).

Table 1. Markers of proliferation and apoptosis (%) in the tested groups; n = 10 in each group; ***—$p < 0.001$; **—$p < 0.01$—between rat glioma without and after PBM; †††—$p < 0.001$; ††—$p < 0.01$ between PBM during sleep or wakefulness; n = 10 in each group; the ANOVA test with post hoc Duncan test.

Markers	Control (Healthy Rats)	Control (Sham Rats)	Control (Sham Rats PBM_Sleep)	Control (Sham Rats PBM_Awake)	Rats with Glioma, No PBM	Rats with Glioma, PBM_Awake	Rats with Glioma, PBM_Sleep
Ki67	-	-	-	-	98.87 ± 1.1	27.44 ± 6.0 ***	11.69 ± 1.0 ***†††
Bax	-	-	-	-	16.29 ± 4.3	65.33 ± 7.5 **	96.91 ± 3.8 ***††
p53	-	-	-	-	20.25 ± 6.1	47.00 ± 4.4 **	87.33 ± 7.4 ***†††

The data of the IHC study show that the expression of the marker of proliferation (Ki67) was higher in rats with glioma and without the PBM course vs. rats treated by the PBM course during the sleep or awake state. However, the proliferation of glioma cells was suppressed more strongly in rats that received the PBM course during the sleep vs. the awake state.

The expression of the marker of apoptosis Bax, with p53 as its regulating factor, was lower in glioma without the PBM course vs. the PBM course during the sleep and awake states (Table 1). Notably, activation of apoptosis in glioma cells was significantly higher in rats receiving the PBM course during the sleep compared to the awake state.

Thus, both PBM courses during sleep or wakefulness effectively increased resistance to the progression of glioma via suppression of proliferation of the tumor cells and activation of apoptosis. Interestingly, the effects of the PBM course on the expression of Ki67, Bax, and p53 were 2.3-, 1.5-, and 1.8-fold greater in rats receiving the PBM course during the sleep vs. the awake state (Table 1).

3.3. The Effects of the PBM Course during Sleep or Awake State on BD

Deep sleep is accompanied by an increase in the size of the perivascular spaces, which facilitates the movement of brain fluids and promotes activation of BD [19]. Since in our previous study on awake rats we found that PBM can stimulate BD, here we tested our hypothesis that BD will be more strongly activated by PBM in sleeping vs. awake rats due to a PBM-related enhancement in natural BD activation.

For this purpose, the lymphatic removal of FITCD from the brain to the dcLNs was studied. The results of ex vivo confocal imaging of the whole brain revealed a significant reduction in FITCD spreading in brain tissues and its lymphatic removal to the dcLNs in rats with glioma without the PBM course vs. the control, including healthy rats (Figure 3a,b,e,f,i,j).

Indeed, the quantitative analysis found that the intensity of the fluorescent signal from FITCD in the ventral and dorsal parts of the brain, as well as in the dcLNs, was 7.5-, 9.3-, and 5.1-fold greater in healthy rats vs. rats with glioma, respectively (Figure 3m–o). Both PBM courses during sleep or wakefulness essentially improved the FITCD distribution in brain tissues and its further accumulation in the dcLNs (Figure 3b–d,f–h,j–o). However, the stimulating PBM effects on BD was higher in rats receiving the PBM course during sleep vs. when awake (Figure 3c,d,g,h,k–o). Thus, these findings demonstrate that glioma-mediated suppression of BD can be improved by the PBM course, with more pronounced effects of PBM during the sleep vs. the awake state.

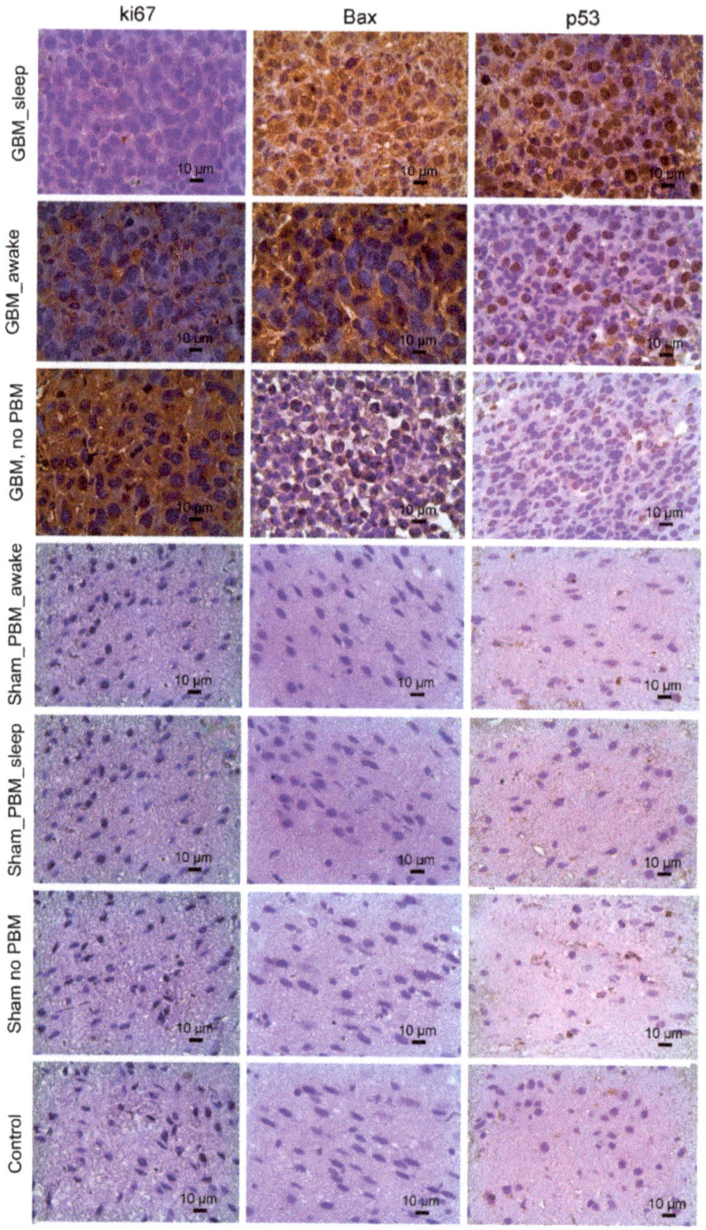

Figure 2. IHC analysis of proliferation and apoptosis of glioma cells in seven tested groups, including control (healthy rats) group; 3 sham groups, in which animals received an injection of saline in the same volume and location as the glioma cells were implanted without PBM and after the PBM course during sleep and awake states; 3 glioma groups, in which rats were with glioma without PBM and after the PBM course during sleep and awake states. The quantitative analysis of immunopositive cells (shown in brown) expressing K67, Bax, and p53 is presented in Table 1.

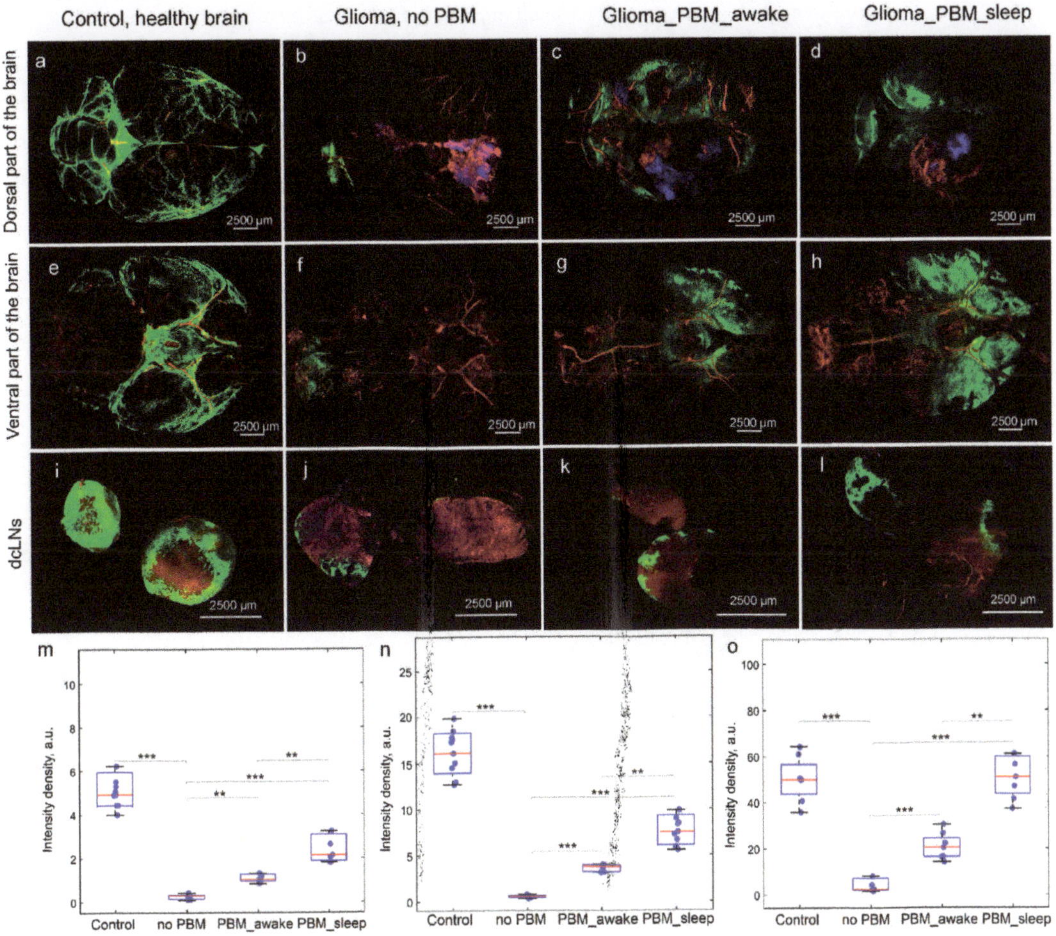

Figure 3. The effects of the PBM course during sleep or wakefulness on BD in the tested groups: (**a–d**) and (**e–h**) Representative images of FITCD spreading in dorsal (**a–d**) and ventral (**e–h**) parts of the brain from the control group (healthy brain) (**a**,**e**), the glioma group without the PBM course (**b**,**f**), and the groups treated by the PBM course during wakefulness (**c**,**g**) or sleep (**d**,**h**); (**i–l**) representative images of accumulation of FITCD in dcLNs in rats from the control group (**i**), the glioma group without the PBM course (**j**), and the groups treated by the PBM course during wakefulness (**k**) or sleep (**l**); (**m–o**) quantitative analysis of intensity of fluorescent signal from FITCD in dorsal (**m**) and ventral (**n**) parts of the brain and in dcLNs (**o**); $n = 10$ in each group; ***—$p < 0.001$; **—$p < 0.01$ between groups; the ANOVA test with post hoc Duncan test.

3.4. The Effects of the PBM Course during Sleep or Wakefulness on Brain's Tumor Immunity

Accumulating evidence indicates that increasing cytotoxic CD8+ T cell infiltration improves survival in glioma [42]. Indeed, activated CD8+ T cells activate the apoptosis, providing suppression of tumor growth [43–45]. However, the lack of signaling molecules in glioma, which activate the generation of lymphocytes in the lymph nodes, is a main reason for the limitation of CD8+ T cells in the tumor region, leading to immune suppression [46,47]. Therefore, in the next step, we investigated the stimulating effects of the PBM course during the sleep or awake state on the immune response against glioma.

Figure 4a,d,g,h demonstrate only a small number of CD8+ T cells around the fluorescent glioma and in the dcLNs in rats with glioma and without the PBM course. Notably, the number of CD8+ T cells in glioma cells and in the dcLNs was significantly increased after the PBM course, both during sleep or wakefulness (Figure 4a–h). However, the PBM course during sleep more strongly increases the number of CD8+ T cells in glioma and in the dcLNs than during awake state (Figure 4b,c,e–h). These data clearly show that the PBM course activates an immune response against the glioma growth that is more effective if PBM is applied during sleep than during wakefulness. Figure 4i schematically illustrates the PBM-mediated stimulation of the generation of CD8+ cells in dcLNs and their extravasation into the microenvironment of the glioma.

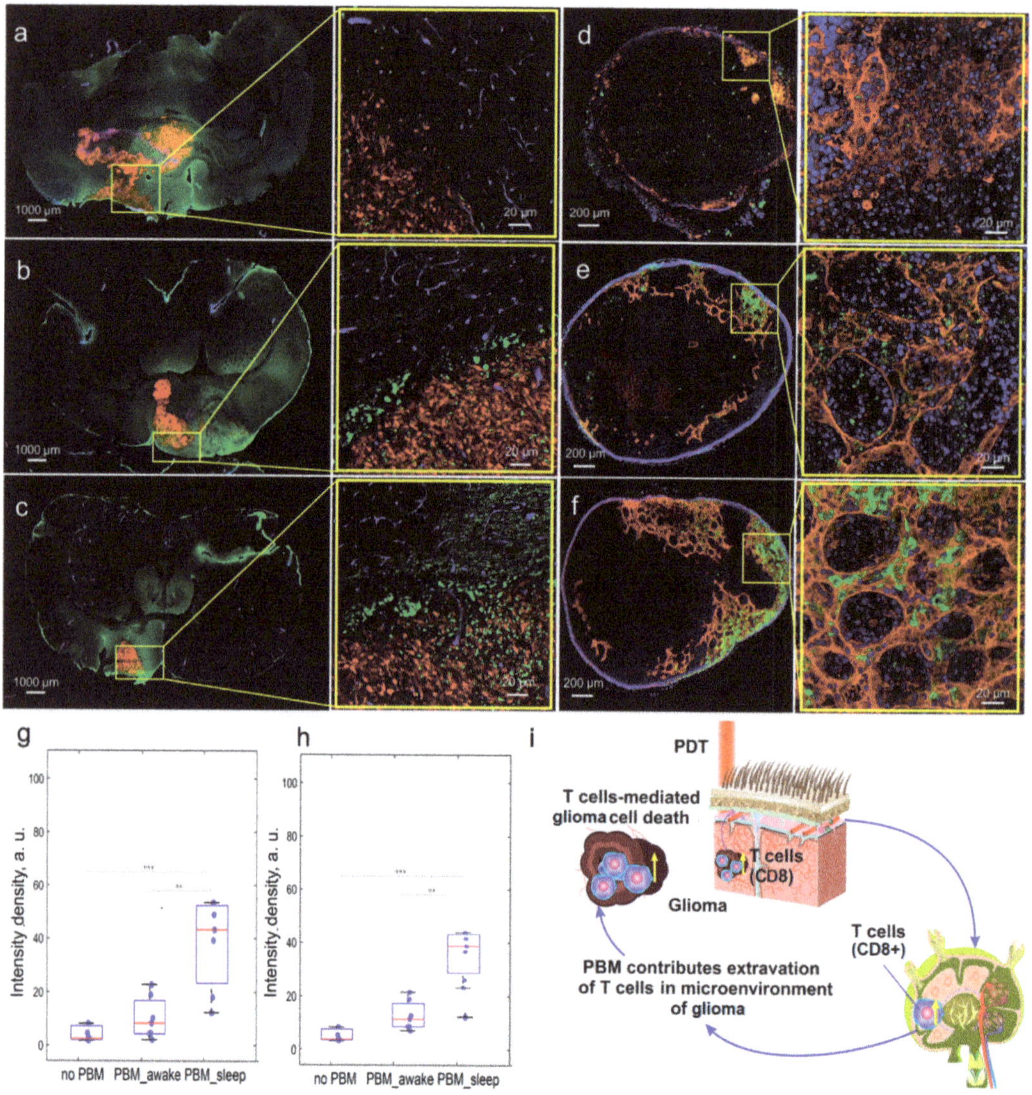

Figure 4. The effects of the PBM course during sleep or wakefulness on brain's tumor immunity: (**a**–**f**) Representative images of fluorescent glioma (red) and CD8+ T cells (green) in the brain (**a**–**c**)

and in dcLNs (**d–f**) in rats with glioma and without the PBM course (**a**,**d**), in rats with glioma receiving the PBM course during wakefulness (**b**,**e**), and in rats with glioma and receiving the PBM course during sleep (**c**,**f**); in (**d–f**): CD8+ T cells (green), LVs (red), and DAPI (blue); (**g**,**h**)—quantitative analysis of intensity of fluorescent signal from CD8+ T cells in glioma (**g**) and in dcLNs (**h**); $n = 10$ in each group, ***—$p < 0.001$; **—$p < 0.01$ between groups; the ANOVA test with post hoc Duncan test; (**i**)—schematic illustration of the PBM effects on an immune response against the glioma growth via increasing the number of cytotoxic CD8+ cells in glioma cells and in dcLNs.

4. Discussion

In this study, we clearly demonstrate that PBM has a therapeutic effect on rat glioma that is consistent with our previous data [24]. Notably, the PBM course during sleep suppresses glioma progression more strongly and increases survival compared to PBM during the awake state. These results were expected since we assumed that PBM during sleep increases BD, as we also showed in our earlier studies [21,22]. Indeed, we found dramatically reduced BD in rats with glioma that was significantly improved after both the PBM courses during either the sleep or awake state. However, the PBM effects on BD were statistically higher in sleeping vs. awake rats. This can be explained by the fact that sleep is associated with the activation of BD and lymphatic removal of metabolites and toxins from the brain [19–22]. Thus, PBM during sleep affects naturally activated BD. The reduced BD in subjects with glioma is accompanied by a lack of MLVs [16–18]. Therefore, intracisternal injection of the vascular endothelial growth factor C for stimulation of lymphaneogenesis is proposed as a new promising strategy for the treatment of brain tumors [18]. PBM can also improve the MLV functions, as we clearly demonstrated in our previous study using photoablation of MLVs [22]. In this study, we show that PBM-mediated restoration of cleaning and drainage functions of MLVs after injuries is better after PBM during sleep vs. when awake [22]. Interestingly, the PBM course during sleep vs. when awake promotes better learning in mice [35].

The mechanisms of PBM-suppression of glioma and the improvement of survival could be due to PBM-mediated activation of apoptosis and blockage of the proliferation of glioma cells. These effects of PBM were also more pronounced in sleeping vs. awake rats. These data are confirmed by our previous findings of the cellular mechanisms of phototherapy for glioma [24]. The application of PBM during sleep-associated activation of BD can contribute to the optimization of immune responses against glioma due to improvements in the microenvironment of the tumor and compensation of the CSF outflow. Indeed, we revealed a lack of CD8+ T cells in both brain tumors and in dcLNs in rats with glioma that has also been shown in other studies [46,47]. However, PBM significantly improved the number of tumor-infiltrating cytotoxic CD8+ T cells more significantly in sleeping vs. awake rats. Nakano et al. report that an increased proliferative activity of CD8+ T cells associated with a decrease in the Ki67 expression is a favorable prognostic factor for survival in patients with cancer [48]. CD8+ T cells can produce antitumor antibodies, support antitumor immune responses, and help to prime immune responses at the tumor site [44,49].

Currently, photodynamic therapy (PDT) is considered as a promising tool as an alternative therapy for glioma [50–55]. PDT combines a light source, which excites a photosensitizer accumulated specifically in the glioma cells. The excited photosensitizer interacts with triplet oxygen and produces singlet oxygen, damaging the tumor cells by direct (necrosis and apoptosis) and indirect (occlusion of tumor vessels and enhanced host immunity) injuries [51]. There are several clinical studies suggesting the therapeutic effects of PDT in patients with glioma using different photosensitizers, as we discussed in our review [51]. However, PDT cannot be used in patients with allergies to photosensitizers, including newborns, due to side effects of the photosensitizers. Indeed, there are disadvantages of PDT, including photosensitivity after treatment, erythema, edema, blisters, urticarial, photophobia, heat-shock response, and an increase in the permeability of the blood–brain barrier [56–60]. Recently, we proposed a new strategy for phototherapy for glioma, using

PBM without photosensitizers in awake animals [24]. There is emerging evidence that sleep can significantly increase the therapeutic properties of PBM [21,22,35]. However, the idea of PDT of brain diseases during sleep is in its infancy [61]. Therefore, there are no technologies for simultaneous PBM and sleep monitoring [61]. In this study, we propose the effective application of our new technology for sleep-phototherapy for glioma that incorporates modern state-of-the-art facilities of optoelectronics and biopotential detection and that can be built of relatively cheap and commercially available components.

5. Conclusions

In this study, using new technology for sleep-phototherapy for glioma we showed for the first time that the therapeutic effects of PBM during sleep in rats with glioma are significantly enhanced compared with the use of PBM during the awake state. Indeed, the PBM course in sleeping rats vs. awake ones suppresses glioma growth more strongly and increases survival compared with the control. A new result was the fact that the PBM course during the sleep vs. awake state has greater effects on the immune response against glioma, including an increase in the number of cytotoxic CD8+ in glioma cells that was associated with activation of apoptosis on the glioma cells and suppression of their proliferation. We also found stronger stimulation of BD by PBM in sleeping vs. awake rats, which is consistent with our findings in previous studies [21,22]. Our new technology for sleep-phototherapy for glioma opens a new strategy to improve the quality of medical care for patients with brain cancer using promising smart-sleep and non-invasive approaches of glioma treatment.

Author Contributions: Conceptualization, A.S. and O.S.-G.; methodology, I.F., E.I., D.M. and D.T.; software, E.I. and I.F.; formal analysis, M.T. (Maria Tzoy) and I.B.; investigation, I.B., A.T., V.A., D.Z., A.E., V.T., A.D. (Alexander Dubrovsky), M.M., V.K., M.T. (Matvey Tuzhilkin), I.E., N.N., E.S., T.I. and E.L.; resources, A.S.; data curation, A.S. and O.S.-G.; writing—original draft preparation, A.S., I.B., A.T., I.F., E.I., D.M. and O.S.-G.; writing—review and editing, A.S. and O.S.-G.; visualization, I.B., A.T., A.D. (Alexander Dubrovsky), A.D. (Alexander Dmitrenko) and I.F.; supervision, A.S. and O.S.-G.; project administration, A.S.; funding acquisition, A.S. All authors have read and agreed to the published version of the manuscript.

Funding: This research was supported by grant from Russian Science Foundation № 23-25-00296.

Institutional Review Board Statement: All experimental procedures were performed in accordance with the "Guide for the Care and Use of Laboratory Animals", Directive 2010/63/EU on the Protection of Animals Used for Scientific Purposes, and the guidelines from the Ministry of Science and High Education of the Russian Federation (№ 742 from 13 November 1984), which have been approved by the Bioethics Commission of the Saratov State University (Protocol No. 8, 18 April 2023).

Informed Consent Statement: Not applicable.

Data Availability Statement: The data that support the findings of this study are available on request from the corresponding author.

Acknowledgments: We thank CCU «Simbioz» and immunochemistry laboratory IBPPM RAS for their support with the immunofluorescence analysis and confocal microscopy within the current research project.

Conflicts of Interest: The authors declare no conflicts of interest.

References

1. Noorvash, B.; Amaral, L.J.; Nelson, T.; Rudnick, J.D.; Irwin, S.A.; Freedland, S.J.; Gresham, G.; Hu, J.L. Remote monitoring of sleep activity in patients diagnosed with glioblastoma. *J. Clin. Oncol.* **2022**, *16*, e14036. [CrossRef]
2. Martin, J.A.; Hart, N.H.; Bradford, N. Prevalence and management of sleep disturbance in adults with primary brain tumours and their caregivers: A systematic review. *J. Neurooncol.* **2023**, *162*, 25–44. [CrossRef] [PubMed]
3. Büttner-Teleagă, A.; Kim, Y.T.; Osel, T.; Richter, K. Sleep disorders in cancer—A systematic review. *Int. J. Env. Res. Public. Health* **2021**, *18*, 11696–11734. [CrossRef] [PubMed]
4. Otte, J.L.; Carpenter, J.S.; Manchanda, S.; Rand, K.L.; Skaar, T.C.; Weaver, M. Systematic review of sleep disorders in cancer patients: Can the prevalence of sleep disorders be ascertained? *Cancer Med.* **2015**, *4*, 183–200. [CrossRef] [PubMed]

5. Jeon, M.S.; Dhillon, H.M.; Agar, M.R. Sleep disturbance of adults with a brain tumor and their family caregivers: A systematic review. *Neuro Oncol.* **2017**, *19*, 1035–1046. [CrossRef] [PubMed]
6. Armstrong, T.S.; Shade, M.Y.; Breton, G.; Gilbert, M.R.; Mahajan, A.; Scheurer, M.E. Sleep-wake disturbance in patients with brain tumors. *Neuro Oncol.* **2016**, *19*, 323–335. [CrossRef]
7. Fox, S.W.; Lyon, D.; Farace, E. Symptom clusters in patients with high-grade glioma. *J. Nurs. Sch.* **2007**, *39*, 61–67. [CrossRef]
8. Huang, Y.; Jiang, Z.-J.; Deng, J.; Qi, Y.-J. Sleep quality of patients with postoperative glioma at home. *World J. Clin. Case* **2020**, *8*, 4735–4742. [CrossRef]
9. Lin, M.R.; Chen, P.Y.; Wang, H.C.; Lin, P.C.; Lee, H.C.; Chiu, H.Y. Prevalence of sleep disturbances and their effects on quality of life in adults with untreated pituitary tumor and meningioma. *J. Neurooncol.* **2021**, *154*, 179–186. [CrossRef]
10. Willis, K.D.; Ravyts, S.G.; Lanoye, A.; Loughan, A.R. Sleep disturbance in primary brain tumor: Prevalence, risk factors, and patient preferences. *Support. Care Cancer* **2022**, *30*, 741–748. [CrossRef]
11. Allgood, J.; Roe, A.; Sparks, B.; Castillo, M.; Cruz, A.; Brooks, A.; Brooks, B. The Correlation of Sleep Disturbance and Location of Glioma Tumors: A Narrative Review. *J. Clin. Med.* **2023**, *12*, 4058. [CrossRef] [PubMed]
12. Sadighi, Z.; Curtis, E.; Zabrowksi, J.; Billups, C.; Gajjar, A.; Khan, R.; Qaddoumi, I. Neurologic impairments from pediatric low-grade glioma by tumor location and timing of diagnosis. *Pediatr. Blood Cancer* **2018**, *65*, 27063. [CrossRef]
13. Subramanian, S.; Gholami, A.; Biros, G. Simulation of glioblastoma growth using a 3D multispecies tumor model with mass effect. *J. Math. Biol.* **2019**, *79*, 941–967. [CrossRef]
14. Gapstur, R.; Gross, C.; Ness, K. Factors associated with sleep-wake disturbances in child and adult survivors of pediatric brain tumors: A review. *Oncol. Nurs. Forum* **2009**, *36*, 723–731. [CrossRef] [PubMed]
15. Berger, A.; Parker, K.; Young-McCaughan, S.; Mallory, G.; Barsevick, A.; Beck, S.; Carpenter, J.; Carter, P.; Farr, L.; Hinds, P.; et al. Sleep wake disturbances in people with cancer and their caregivers: State of the science. *Oncol. Nurs. Forum* **2005**, *32*, E98–E126. [CrossRef] [PubMed]
16. Ma, Q.; Schlegel, F.; Bachmann, S.B. Lymphatic outflow of cerebrospinal fluid is reduced in glioma. *Sci. Rep.* **2019**, *9*, 14815. [CrossRef]
17. Hu, X.; Deng, Q.; Ma, L. Meningeal lymphatic vessels regulate brain tumor drainage and immunity. *Cell Res.* **2020**, *30*, 229–243. [CrossRef]
18. Song, E.; Mao, T.; Dong, H. VEGF-C-driven lymphatic drainage enables immunosurveillance of brain tumours. *Nature* **2020**, *577*, 689–694. [CrossRef]
19. Xie, L.; Kand, H.; Xu, Q. Sleep drives metabolite clearance from the adult brain. *Science* **2013**, *342*, 373–377. [CrossRef]
20. Fultz, N.E.; Bonmassar, G.; Setsompop, K. Coupled electrophysiological, hemodynamic, and cerebrospinal fluid oscillations in human sleep. *Science* **2019**, *366*, 628–631. [CrossRef]
21. Semyachkina-Glushkovskaya, O.; Penzel, T.; Blokhina, I.; Khorovodov, A.; Fedosov, I.; Yu, T.; Karandin, G.; Evsukova, A.; Elovenko, D.; Adushkina, V.; et al. Night photostimulation of clearance of beta-amyloid from mouse brain: New strategies in preventing Alzheimer's disease. *Cells* **2021**, *10*, 3289. [CrossRef] [PubMed]
22. Semyachkina-Glushkovskaya, O.; Shirokov, A.; Blokhina, I.; Fedosov, I.; Terskov, A.; Dubrovsky, A.; Tsoy, M.; Zlatogosrskaya, D.; Adushkina, V.; Evsukova, A.; et al. Mechanisms of phototherapy of Alzheimer's disease during sleep and wakefulness: The role of the meningeal lymphatics. *Front. Optoelectron.* **2023**, *16*, 22.
23. Qin, X.; Liu, R.; Akter, F.; Qin, L.; Xie, Q.; Li, Y.; Qiao, H.; Zhao, W.; Jian, Z.; Liu, R.; et al. Peri-tumoral brain edema associated with glioblastoma correlates with tumor recurrence. *J. Cancer* **2021**, *12*, 2073–2082. [CrossRef]
24. Semyachkina-Glushkovskaya, O.; Sokolovski, S.; Fedosov, I.; Shirokov, A.; Navolokin, N.; Bucharskaya, A.; Blokhina, I.; Terskov, A.; Dubrovski, A.; Telnova, V.; et al. Transcranial photosensitiser-free laser treatment of glioblastoma in rat brain. *Int. J. Mol. Sci.* **2023**, *24*, 13696. [CrossRef] [PubMed]
25. Semyachkina-Glushkovskaya, O.; Abdurashitov, A.; Dubrovsky, A.; Klimova, M.; Agranovich, I.; Terskov, A.; Shirokov, A.; Vinnik, V.; Kuznecova, A.; Lezhnev, N.; et al. Photobiomodulation of lymphatic drainage and clearance: Perspective strategy for augmentation of meningeal lymphatic functions. *Biomed Opt. Express* **2020**, *11*, 725–734. [CrossRef]
26. Semyachkina-Glushkovskaya, O.; Abdurashitov, A.; Klimova, M.; Dubrovsky, A.; Shirokov, A.; Fomin, A.; Terskov, A.; Agranovich, I.; Mamedova, A.; Khorovodov, A.; et al. Photostimulation of cerebral and peripheral lymphatic functions. *Transl. Biophotonics* **2020**, 2, e201900036. [CrossRef]
27. Semyachkina-Glushkovskaya, O.; Fedosov, I.; Shirokov, A.; Vodovozova, E.; Alekseeva, A.; Khorovodov, A.; Blokhina, I.; Terskov, A.; Mamedova, A.; Klimova, M.; et al. Photomodulation of lymphatic delivery of liposomes to the brain bypassing the blood-brain barrier: New perspectives for glioma therapy. *Nanophotonics* **2021**, *12*, 3215–3227. [CrossRef]
28. Li, D.; Liu, S.; Yu, T.-T.; Liu, Z.; Sub, S.; Bragin, D.; Shirokov, A.; Navolokin, N.; Bragina, O.; Zheng-Wu, H.; et al. Photostimulation of brain lymphatics in male newborn and adult rodents for therapy of intraventricular hemorrhage. *Nat. Comm.* **2023**, *14*, 6104. [CrossRef]
29. Bohlen, H.; Gasheva, O.; Zawieja, D. Nitric oxide formation by lymphatic bulb and valves is a major regulatory component of lymphatic pumping. *Am. J. Physiol. Heart Circ. Physiol.* **2011**, *301*, H1897–H1906. [CrossRef]
30. Bohlen, H.; Wang, W.; Gashev, A.; Gasheva, O.; Zawieja, D. Phasic contractions of rat mesenteric lymphatics increase basal and phasic nitric oxide generation in vivo. *Am. J. Physiol. Heart Circ. Physiol.* **2009**, *297*, H1319–H1328. [CrossRef]

31. Kunert, C.; Baish, J.W.; Liao, S.; Padera, T.P.; Munn, L.L. Mechanobiological oscillators control lymph flow. *Proc. Natl. Acad. Sci. USA* **2015**, *112*, 10938–10943. [CrossRef]
32. Petrelli, F.; De Stefani, A.; Ghidini, A.; Bruschieri, L.; Riboldi, V.; Dottorini, L.; Iaculli, A.; Zaniboni, A.; Trevisan, F. Steroids use and survival in patients with glioblastoma multiforme: A pooled analysis. *J. Neurol.* **2021**, *268*, 440–447. [CrossRef] [PubMed]
33. Zoccarato, M.; Nardetto, L.; Basile, A.M.; Giometto, B.; Zagonel, V.; Lombardi, G. Seizures, Edema, Thrombosis, and Hemorrhages: An Update Review on the Medical Management of Gliomas. *Front. Oncol.* **2021**, *11*, 617966. [CrossRef]
34. Ohmura, K.; Tomita, H.; Hara, A. Peritumoral Edema in Gliomas: A Review of Mechanisms and Management. *Biomedicines* **2023**, *11*, 2731. [CrossRef] [PubMed]
35. Semyachkina-Glushkovskaya, O.; Fedosov, I.; Zaikin, A.; Ageev, V.; Ilyukov, E.; Myagkov, D.; Tuktarov, D.; Blokhina, I.; Shirokov, A.; Terskov, A.; et al. Technology of the photobiostimulation of the brain's drainage system during sleep for improvement of learning and memory in male mice. *Biomed. Opt. Express* **2024**, *15*, 44–58. [CrossRef]
36. Hablitz, L.M.; Vinitsky, H.S.; Sun, Q. Increased glymphatic influx is correlated with high EEG delta power and low heart rate in mice under anesthesia. *Sci. Adv.* **2019**, *5*, eaav5447. [CrossRef]
37. Iber, C. *The AASM Manual for the Scoring of Sleep and Associated Events: Rules, Terminology and Technical Specifications*, 1st ed.; American Academy of Sleep Medicine: Darien, IL, USA, 2007.
38. Devos, S.L.; Miller, T.M. Direct intraventricular delivery of drugs to the rodent central nervous system. *J. Vis. Exp.* **2013**, *12*, e50326.
39. Ahn, J.H.; Cho, H.; Kim, J. Meningeal lymphatic vessels at the skull base drain cerebrospinal fluid. *Nature* **2019**, *572*, 62–66. [CrossRef] [PubMed]
40. Schindelin, J.; Arganda-Carreras, I.; Frise, E. Fiji: An open-source platform for biological-image analysis. *Nat. Methods* **2012**, *9*, 676–682. [CrossRef]
41. Bragin, D.; Kameneva, M.; Bragina, O.; Thomson, S.; Statom, G.; Lara, D.; Yang, Y.; Nemoto, E. Rheological effects of drag-reducing polymers improve cerebral blood flow and oxygenation after traumatic brain injury in rats. *J. Cereb. Blood Flow. Metab.* **2017**, *37*, 762–775. [CrossRef]
42. Mauldin, I.S.; Jo, J.; Wages, N.A.; Yogendran, L.V.; Mahmutovic, A.; Young, S.J.; Lopes, M.B.; Slingluff, C.L.; Erickson, L.D.; Fadul, C.E. Proliferating CD8+ T cell infiltrates are associated with improved survival in glioblastoma. *Cells* **2021**, *10*, 3378. [CrossRef]
43. Ferrantelli, F.; Manfredil, F.; Chiozzini, C.; Leone, P.; Giovannelli, A.; Olivetta, E.; Federico, M. Long-term antitumor CD8+ T cell immunity induced by endogenously engineered extracellular vesicles. *Cancers* **2021**, *13*, 2263. [CrossRef] [PubMed]
44. Raskov, H.; Orhan, A.; Christensen, J.P. Cytotoxic CD8+ T cells in cancer and cancer immunotherapy. *Br. J. Cancer* **2021**, *124*, 359–367. [CrossRef] [PubMed]
45. Philip, M.; Schietinger, A. CD8+ T cell differentiation and dysfunction in cancer. *Nat. Rev. Immunol.* **2022**, *22*, 209–223. [CrossRef] [PubMed]
46. Waldman, A.D.; Fritz, J.M.; Lenardo, M.J. A guide to cancer immunotherapy: From T cell basic science to clinical practice. *Nat. Rev. Immunol.* **2020**, *20*, 651–668. [CrossRef] [PubMed]
47. Xia, A.; Zhang, Y.; Xu, J.; Yin, T.; Lu, X.J. T cell dysfunction in cancer immunity and immunotherapy. *Front. Immunol.* **2019**, *10*, 1719. [CrossRef]
48. Nakano, O.; Sato, M.; Naito, Y.; Suzuki, K.; Orikasa, S.; Aizawa, M.; Suzuki, Y.; Shintaku, I.; Nagura, H.; Ohtani, H. Proliferative activity of intratumoral CD8+ T-lymphocytes as a prognostic factor in human renal cell carcinoma: Clinicopathologic demonstration of antitumor immunity. *Cancer Res.* **2001**, *61*, 5132–5136.
49. Xie, Q.; Ding, J.; Chen, Y. Role of CD8+ T lymphocyte cells: Interplay with stromal cells in tumor microenvironment. *Acta Pharm. Sin. B* **2021**, *11*, 1365–1378. [CrossRef]
50. Domka, W.; Bartusik-Aebisher, D.; Rudy, I.; Dynarowicz, K.; Pięta, K.; Aebisher, D. Photodynamic therapy in brain cancer: Mechanisms, clinical and preclinical studies and therapeutic challenges. *Front. Chem.* **2023**, *11*, 1250621. [CrossRef]
51. Abdurashitov, A.; Tuchin, V.; Semyachkina-Glushkovskaya, O. Photodynamic therapy of brain tumors and novel optical coherence tomography strategies for in vivo monitoring of cerebral fluid dynamics. *J. Innov. Opt. Health Sci.* **2020**, *13*, 2030004. [CrossRef]
52. Bartusik-Aebisher, D.; Żołyniak, A.; Barnaś, E.; Machorowska-Pieniążek, A.; Oleś, P.; Kawczyk-Krupka, A.; Aebisher, D. The Use of Photodynamic Therapy in the Treatment of Brain Tumors-A Review of the Literature. *Molecules* **2022**, *27*, 6847. [CrossRef] [PubMed]
53. Stepp, H.; Beck, T.; Pongraz, T.; Meinel, F.; Kreth, W.; Tonn, J.; Stummer, W. ALA and malignant glioma: Fluorescence-guided resection and photodynamic treatment. *J. Environ. Pathol. Toxicol. Oncol.* **2007**, *21*, 157–164. [CrossRef] [PubMed]
54. Mahmoudi, K.; Garvey, K.; Bouras, A.; Cramer, G.; Stepp, H.; Jesu Raj, J.; Bozec, D.; Busch, T.; Hadjipanayis, C. 5-aminolevulinic acid photodynamic therapy for the treatment of high-grade gliomas. *J. Neurooncol.* **2019**, *141*, 595–607. [CrossRef] [PubMed]
55. Correia, J.; Rodrigues, J.; Pimenta, S.; Dong, T.; Yang, Z. Photodynamic Therapy Review: Principles, Photosensitizers, Applications, and Future Directions. *Pharmaceutics* **2021**, *13*, 1332. [CrossRef]
56. Li, Y.; Zhang, P.; Xie, Y.; Yang, J.; Yang, Y.; Shi, L.; Wu, W.; Li, Z. Photosensitizers with multiple degradation modes for efficient and postoperatively safe photodynamic therapy. *Biomaterials* **2023**, *299*, 122182. [CrossRef] [PubMed]
57. Li, X.; Lovell, J.; Yoon, J.; Chen, X. Clinical development and potential of photothermal and photodynamic therapies for cancer. *Nat. Rev. Clin. Oncol.* **2020**, *17*, 657–674. [CrossRef] [PubMed]

58. Wojewoda, K.; Gillstedt, M.; Tovi, J.; Salah, L.; Wennberg Larko, A.; Sjoholm, A.; Sandberg, C. Optimizing treatment of acne with photodynamic therapy (PDT) to achieve long-term remission and reduce side effects. A prospective randomized controlled trial. *J. Photochem. Photobiol. B* **2021**, *223*, 112299. [CrossRef]
59. Ibbotson, S.; Wong, T.; Morton, C.; Collier, N.; Haylett, A.; McKenna, K.; Mallipeddi, R.; Moseley, H.; Rhodes, L.E.; Seukeran, D.C.; et al. Adverse effects of topical photodynamic therapy: A consensus review and approach to management. *Br. J. Dermatol.* **2019**, *180*, 715–729. [CrossRef]
60. Morton, C.; Szeimies, R.; Basset-Seguin, N.; Calzavara-Pinton, P.; Gilaberte, Y.; Haedersdal, M.; Hofbauer, G.F.L.; Hunger, R.E.; Karrer, S.; Piaserico, S.; et al. European dermatology forum guidelines on topical photodynamic therapy 2019 part 2: Emerging indications—Field cancerization, photorejuvenation and inflammatory/infective dermatoses. *J. Eur. Acad. Dermatol. Venereol.* **2020**, *34*, 17–29. [CrossRef]
61. Semyachkina-Glushkovskaya, O.; Fedosov, I.; Penzel, T.; Li, D.; Yu, T.; Telnova, V.; Kaybeleva, E.; Saranceva, E.; Terskov, A.; Khorovodov, A.; et al. Brain Waste Removal System and Sleep: Photobiomodulation as an Innovative Strategy for Night Therapy of Brain Diseases. *Int. J. Mol. Sci.* **2023**, *24*, 3221. [CrossRef]

biomedicines

MDPI

Article

Post-Operational Photodynamic Therapy of the Tumor Bed: Comparative Analysis for Cold Knife and Laser Scalpel Resection

Maria Shakhova [1], Vadim Elagin [2,*], Anton Plekhanov [2], Aleksandr Khilov [3], Daria Kurakina [3], Vladislav Kamensky [2,3] and Mikhail Kirillin [3]

1 Department of Ear, Nose and Throat Diseases, FSBEI HE «Privolzhsky Research Medical University» MOH Russia, 10/1 Minin and Pozharsky Square, Nizhny Novgorod 603005, Russia; maha-shakh@yandex.ru

2 Institute of Experimental Oncology and Biomedical Technologies, FSBEI HE «Privolzhsky Research Medical University» MOH Russia, 10/1 Minin and Pozharsky Square, Nizhny Novgorod 603005, Russia; strike_gor@mail.ru (A.P.); vlad@ufp.appl.sci-nnov.ru (V.K.)

3 A.V. Gaponov-Grekhov Institute of Applied Physics of the Russian Academy of Sciences, 46 Ulyanov St., Nizhny Novgorod 603155, Russia; alhil@inbox.ru (A.K.); vekfy@inbox.ru (D.K.); mkirillin@yandex.ru (M.K.)

* Correspondence: elagin.vadim@gmail.com; Tel.: +7-83-1465-5672

Abstract: In this paper, we report on a study regarding the efficiency of the post-operational phototherapy of the tumor bed after resection with both a cold knife and a laser scalpel in laboratory mice with CT-26 tumors. Post-operational processing included photodynamic therapy (PDT) with a topically applied chlorin-based photosensitizer (PS), performed at wavelengths of 405 or 660 nm, with a total dose of 150 J/cm^2. The selected design of the tumor model yielded zero recurrence in the laser scalpel group and 92% recurrence in the cold knife group without post-processing, confirming the efficiency of the laser scalpel in oncology against the cold knife. The application of PDT after the cold knife resection decreased the recurrence rate to 70% and 42% for the 405 nm and 660 nm procedures, respectively. On the other hand, the application of PDT after the laser scalpel resection induced recurrence rates of 18% and 30%, respectively, for the considered PDT performance wavelengths. The control of the penetration of PS into the tumor bed by fluorescence confocal microscopy indicated the deeper penetration of PS in the case of the cold knife, which presumably provided deeper PDT action, while the low-dose light exposure of deeper tissues without PS, presumably, stimulated tumor recurrence, which was also confirmed by the differences in the recurrence rate in the 405 and 660 nm groups. Irradiation-only light exposures, in all cases, demonstrated higher recurrence rates compared to the corresponding PDT cases. Thus, the PDT processing of the tumor bed after resection could only be recommended for the cold knife treatment and not for the laser scalpel resection, where it could induce tumor recurrence.

Keywords: photodynamic therapy; tumor resection; laser surgery; cold knife; chlorin-based photosensitizers; Monte Carlo simulations

1. Introduction

The introduction of laser scalpels into oncological surgery in recent decades [1,2] was a significant advance, ensuring R0 resection, owing to their efficient cutting capabilities combined with tissue carbonization, a considerable reduction in bleeding, minimized intraoperative trauma, and the absence of coarse scars in the postoperative period. Diode lasers have been shown to reduce tumor dissemination during surgical procedures due to bleeding prevention [3,4]. However, postoperative local recurrences still remain an important problem in the clinical treatment of cancers. In this connection, the additional improvement in surgical protocols to additionally decrease the risk of further tumor growth, and the probability of recurrence is of high importance [5]. In particular, a protocol including a procedure allowing for the elimination of cancer cells left in tissue after surgical intervention could be advantageous against traditional approaches. A combined approach to the treatment of tumors of different localizations may include photodynamic therapy (PDT) [6–8], antitumor immunity [9,10], cryotherapy [11–13], hyperthermia [14,15], and radiotherapy [16–18], in addition to surgical treatment, which helps to reduce the recurrence rate, and this can not only increase the survival period but also improve the quality of life of cancer patients [19]. In addition, considering the possibility of invasive tumor growth and the spread of tumor cells beyond the visually defined resection margin, as well as the fact that surgical intervention itself does not provide ablasticity, improvements in both the surgical protocols and the introduction of additional intraoperative treatment methods would ensure R0 resection is relevant.

Photodynamic therapy is a low-invasive modern treatment technique based on the photoactivation of a photosensitizer, resulting in the production of singlet oxygen and other reactive oxygen species, and which has proved to be efficient in oncology [20,21]. This technique does not require accurately targeting cancer cells and provides action to the areas with an accumulated photosensitizer (PS) irradiated with light of a specific wavelength. Thus, the configuration of the irradiation beam determines the shape of the treated volume. The PDT intraoperatively used for the treatment of the tumor bed to control the remaining cells of the primary tumor and reduce the risk of recurrence was first described in the 1990s [22]. When performing PDT in oncology, the intravenous administration of a photosensitizer is traditionally used. It should be noted that even a minimal accumulation of PS in the skin can trigger a photochemical reaction under the influence of daylight and lead to the development of complications after the PDT procedure in the absence of preventive measures. These limitations, as well as the economic and time aspects of the PDT procedure, have led to the development of topical formulations of PS with more favorable photophysical properties in the form of ointments, solutions, and gels.

Topically applied PSs do not require intravenous injection, and they benefit from a simple application procedure [23]. Chlorin-based photosensitizers feature two peaks in their absorption spectrum in the red and blue bands, thus providing different penetration depths owing to significant differences in the biotissue optical properties of these bands [24,25]. Previously, PDT with topically applied PS has demonstrated limited efficiency in tumor treatment, owing to the limited penetration of PS into the tumor [26]; however, its application to the tumor bed after the procedure seems to be promising. The choice of blue or red light for irradiation provides the opportunity for performing strong superficial action or a deeper action with weaker depth dependence based on the evaluation of the tumor cell invasion depth. Numerical Monte Carlo simulations have demonstrated significant differences in the in-depth distribution of the absorbed light dose governing PDT treatment areas for blue and red bands [25].

The introduction of PDT gels [27] has raised the question regarding the additional treatment of the surgical tumor bed during organ-preserving operations [28,29]. In this study, we compared the result of using PDT after the local application of the PS gel after tumor excision using a cold knife and a contact laser scalpel with an optical fiber that has the shape of a knife. The theory of "correct geometry" and the development of the methodology of the contact laser scalpel application have recently been developed [30]. An important feature is the contact laser cutting, which seals blood vessels and results in the absence of bleeding, as well as the minimization of the tumor cell spread into healthy tissue when working in ablative mode. The main criterion for treatment efficiency, including the surgical removal of the tumor, is the absence of continued growth and recurrences over long periods, with an assessment of the five-year survival rate. Different coagulation properties of the wound surface after contact laser surgery and cold knife tumor resection, in particular, the different diffusions of the PDT gel into the biological tissue, stimulate the comparison of a combination of these techniques with PDT for the post-operative elimination of the remaining cancer cells.

The aim of this study was to compare the effect of additional PDT of the tumor bed with topical PS application after using a cold knife (no coagulation layer) and after contact excision with a waveguide laser scalpel operating at 0.97 μm, forming a coagulation layer with a thickness of about 300 μm [31]. This study was performed with a chlorin-based PS in order to additionally compare the effects of the PDT procedure at 405 and 660 nm. The invasion of the employed tumor model into the muscular tissues beyond the tumor node was verified by histology. The distribution of the PS in tissue upon administration was analyzed using fluorescence microscopy, while the distribution of the absorbed light dose at the considered wavelengths was studied using Monte Carlo simulations.

2. Materials and Methods

2.1. Animal Cohort

All animal studies were approved by the Ethics Committee of the Privolzhsky Research Medical University (Protocol No. 13 of 7 July 2021). Experiments were performed on 113 female Balb/c mice. Mice were subcutaneously inoculated in the left flank with mouse colon adenocarcinoma (CT-26) cells (in an amount of 2.5×10^5 cells per injection) in 100 μL of phosphate-buffered saline. All animals were blindly divided into 10 experimental groups according to the type of post-operational treatment of the tumor bed (Table 1). The perioperative death cases during the experiment in different groups occurred within 2 h after the procedure and were caused by the complications of the narcotization procedure.

Table 1. Groups of experimental animals depending on the type of resection and post-operative treatment of the tumor bed (numbers in parentheses show perioperative deaths).

Group No.	Resection Method	Treatment of the Tumor Bed	Number of Animals
1	laser scalpel	Irradiation only @660 nm	11 (3)
2	laser scalpel	PDT: Radagel® @660 nm	11 (1)
3	laser scalpel	Irradiation only @405 nm	11 (-)
4	laser scalpel	PDT: Radagel® @405 nm	11 (1)
5	laser scalpel	No additional treatment	11 (-)
6	cold knife	Irradiation only @660 nm	11 (2)
7	cold knife	PDT: Radagel® @660 nm	11 (-)
8	cold knife	Irradiation only @405 nm	10 (-)
9	cold knife	PDT: Radagel® @405 nm	11 (1)
10	cold knife	No additional treatment	13 (1)
	Total		113 (9)

2.2. Resection Procedure

Tumors were resected on the 10th day after inoculation, when the tumor node volume had reached approximately 1 cm^3. The mice were anesthetized intramuscularly with a mixture of Zoletil (40 mg/kg, Virbac, Carros, France) and 2% Rometar (10 mg/kg, Spofa, Jičín, Czech Republic) before the tumor node resection. The skin on the tumor site was cut using a cold knife. The tumor nodes were resected using either a commonly used cold knife or a laser scalpel. An LSP-0.97/10 (IRE-Polyus, Fryazino, Russia) laser scalpel operating at a wavelength of 0.97 μm with an output power of 6 W was used to excise the tumor. The laser light was delivered to the tissue via a silica fiber of 550 μm in diameter. The surgeon removed the tumor node in accordance with visual evaluation of tumor margins.

2.3. PDT Procedure

To perform a photodynamic therapy procedure, a chlorin e6-containing gel (Radagel®, "RADA-PHARMA" Ltd., Moscow, Russia) was used. The gel contains the sodium salt of chlorin e6 as an active substance in concentration of 0.0175 mg per 10 mL. Approximately 100 mg of the gel was topically applied onto the surface of the tumor bed by a cotton swab and was evenly distributed over the surgical field. The gel was kept for 20 min and then removed with wet gauze pads. In the course of the PDT procedure of the tumor bed after the tumor excision, the irradiation was performed with the PDT device "Harmonia" (Laser MedCenter Ltd., Moscow, Russia) with a total dose of 150 J/cm^2; the fluence rate at the tissue surface was 260 and 125 mW/cm^2 for the wavelengths of 660 and 405 nm, respectively. The irradiation spot size was 9 mm in diameter. The fluence rate was different for 660 and 405 nm since our previous studies indicated that the equal fluence rate results in extensive heating of the tumor site in course of irradiation at a wavelength of 405 nm. On the other hand, we wanted to minimize the procedure time and limit the irradiation procedure to within half an hour to prevent mice from waking up from narcosis during irradiation. The irradiation-only treatment was performed in the same conditions, with the same device and doses, as the PDT treatment, though without application of the PS gel.

2.4. Follow-Up

The appearance of tumor recurrence was monitored for 3 months after the resection. Upon completion of the monitoring or detection of tumor recurrence reaching 1 cm^3 in size, the animals were sacrificed. All animal procedures followed the Guidelines for Works Involving Experimental Animals and the International Guiding Principles for Biomedical Research Involving Animals and the ethical principles established by the European Convention for the Protection of Vertebrate Animals used for Experimental and other Scientific Purposes.

For comparative analysis, we tried to standardize the experiment; the resection was carried out by one surgeon, the mice were from the same series, and the PDT procedure was carried out with the same device and personnel.

2.5. Analysis of Photosensitizer Distribution

A study was performed in order to analyze the penetration of the PS into the tumor bed upon topical application. The skin above the tumor node and the part of the tumor of about 3 mm in thickness were removed using a cold knife or a laser scalpel. The study involved 5 animals for each resection technique. The Radagel PS was applied onto the obtained surface and the animals were kept in a dark box. After 20 min, the PS was removed from the surface and the tumors were excised. The excised tumors were embedded in O.C.T. compound and immersed in ethanol cooled to −80 °C. Twenty-micrometer-thick cryosections oriented normal to the tissue surface were prepared using a bench-top cryostat CM 1100 (Leica Biosystems, Nussloch, Germany) at a temperature of −20 °C. To avoid redistribution of the PS on the slice surface, the tumor samples were oriented in such a way that the cryostat knife reached the surface of the Radagel administration after passing through the entire sample. Fluorescence intensity of cryosections was studied using laser-scanning microscope LSM 880 (Carl Zeiss, Jena, Germany). An oil-immersion objective C-Apochromat 40×/1.45 NA was used for image acquisition. Chlorin e6 fluorescence was excited at a wavelength of 633 nm with a He-Ne laser. Fluorescence emission was detected in the band of 650–735 nm. For each sample, 12 tile scans from 4 cryosections were acquired. The tile scan consists of 6 serial images captured consequently, starting from the sample surface. The fluorescence intensity images were processed with ZEN 3.0 (Carl Zeiss) and ImageJ 1.39p software (NIH, Bethesda, MD, USA).

2.6. Monte Carlo Simulations

In order to interpret the results of the experiment, we performed a simulation of the distribution of the absorbed dose within tissue at two employed wavelengths. Simulations of the absorbed light distribution were conducted with the use of our previously reported code [32] for Monte Carlo modeling of light transport in biological tissues. To analyze the absorbed light distribution, models of tumor and muscle mouse tissues were considered. The simulations were performed for uniform illumination of the tissue sample at wavelengths of 405 and 660 nm. It is worth noting that the optical properties data reported for mouse tumor models as well as mouse muscle tissues are quite limited. Absorption and reduced scattering coefficients for tumor tissue model were chosen in accordance with [33,34]. Due to the lack of literature data, optical properties of mouse muscle tissue were reconstructed from our ex vivo spectrophotometric measurements [35] using the look-up-table method described in [36]. The considered optical properties for the wavelengths of 405 and 660 nm are summarized in Table 2.

Table 2. Absorption μ_a and reduced scattering μ_s' coefficients and reflective index n used in Monte Carlo simulations.

Wavelength, nm	μ_a, mm^{-1}	μ_s', mm^{-1}	n
		Tumor tissue	
405 nm	0.92	1.29	1.4
660 nm	0.18	0.67	1.4
		Muscle tissue	
405 nm	0.57	1.17	1.4
660 nm	0.07	0.58	1.4

For the considered optical properties of the tissue model, a 3D map of the absorbed light dose $A(x,y,z,\lambda)$ with a wavelength λ was calculated and transformed into the in-depth density of the absorbed dose fraction:

$$\tilde{A}(z, \lambda) = \iint \frac{A(x,y,z,\lambda)}{dVN_{ph}} dx\, dy$$

normalized to the value of $N_{ph}dV$ [32], where $N_{ph} = 10^7$ is the total number of launched photons, and dV is the volume of a voxel.

3. Results and Discussion

3.1. Histology Study

Since the aim of this study was to compare the efficiency of different treatment protocols in suppressing tumor growth and preventing recurrence, it was important to choose the model with the high probability of tumor invasion into muscular tissue, when accurate surgical removal of tumor node with a cold knife may not ensure R0 resection [37]. In order to evaluate the adequacy of the chosen model, we performed a histological study of a typical tumor on the 10th day after inoculation. An image of a hematoxylin–eosin (H&E)-stained tissue sample showing the tumor node and underlying muscular tissue is shown in Figure 1. This figure demonstrates the invasion of tumor cells into the layer of muscular tissue beneath the tumor node. Two arbitrary areas within the muscular layer shown with higher magnification clearly demonstrate tumor cells located between muscle fiber bundles as well as disorganization of the bundle structure owing to tumor proliferation. The observed approval of the high tumor cell count outside the tumor node indicates the adequacy of the chosen model with respect to the aims of the study. Note that the tumor cells could be found as deep as 1 mm below the tumor node margin.

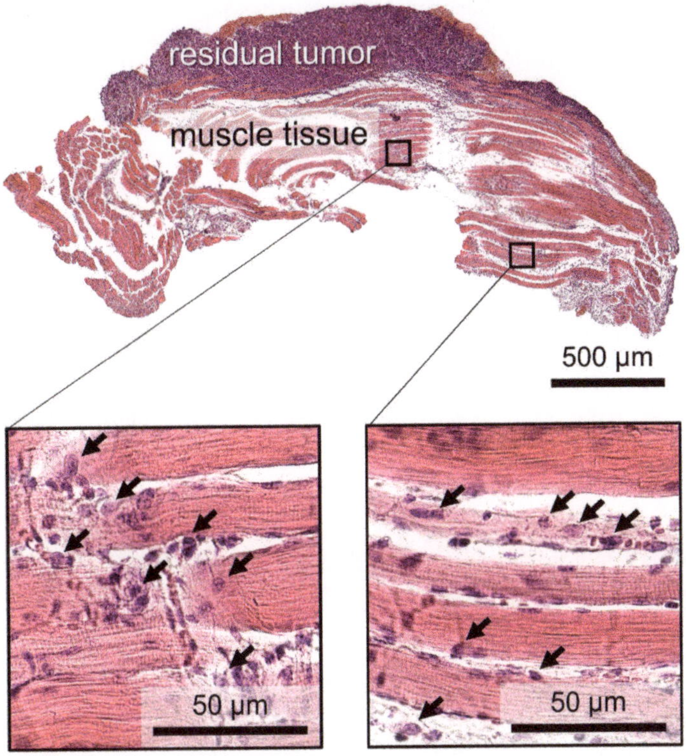

Figure 1. Histological image of the tumor bed after removal with a cold knife. The arrows show revealed tumor cells.

3.2. Fluorescence Microscopy

In our previous study, we demonstrated a high efficiency of employing a laser scalpel instead of a cold knife in preventing tumor recurrence [38], while the aim of this study was to evaluate the additional effect of a complementary PDT procedure on the tumor bed after surgical treatment. As demonstrated earlier, the use of a laser scalpel results in the formation of a caramelization layer [39] on the tissue surface, the parameters of which are described in other papers [30,31]. The studies [30,31] demonstrated that the thickness of this layer is usually about 300–400 µm. One may expect the difference in the penetration of a topically applied PS into tissue after resection with the cold knife and the laser scalpel owing to the differences in the properties of this layer. In order to demonstrate this difference, an analysis of the PS distribution upon topical application was performed using fluorescence microscopy images of tumor cross-sections stitched from several fields of view. Figure 2 shows typical fluorescence microscopy images, corresponding to the cold knife (Figure 2a) and laser scalpel (Figure 2b) tumor resection. One can see that the PS penetrates much deeper into tissue in the case of the cold knife resection. The cold knife case demonstrates a pronounced fluorescence signal to depths of about 600 µm, while for the laser scalpel case, this depth is below 200 µm.

In order to quantify the difference in in-depth PS distribution for the two considered surgical options, we plotted the dependencies of the integral fluorescence signal in an image tile on depth averaged over five independent images each. The results of the fluorescent PS in-depth distribution analysis are shown in Figure 2c on the logarithmic scale. The observed dependencies confirm the exponential in-depth decay of the PS concentration for both cases, which is in agreement with the predictions of the analytical model of substance diffusion into tissue [40]. Comparisons of the two dependencies indicate both a lower fluorescence signal from the laser scalpel-cut tissue and faster in-depth attenuation of the PS concentration. The thickness of the caramelization layer was estimated previously, amounting to about 400 μm, so the dependence shows that the fluorescence signal from the depth of 636 μm can be assumed to correspond to a depth below the caramelization layers that is more than 20-times smaller compared to the signal from the superficial layer. At the same time, the corresponding attenuation of the fluorescence signal for the cold knife is less than 7-times. The observed dependencies indicate the difference in the in-depth PS concentration distribution between two considered cases, so one can expect deeper PDT action for the cold knife option. On the other hand, it is worth noting that the concentration reaches an asymptotic nature at a depth of about 1 mm, allowing us to assume that, at least in some cases, the concentration of PS at this depth is negligible. This fact allows one to expect that the PDT action at this depth could be insufficient to suppress tumor cells. Several studies have previously demonstrated that low-dose PDT [41] or topical administration of PS [26] may result in an incomplete response of a tumor to treatment. Moreover, it was demonstrated earlier that low-dose PDT may provoke tumor recurrence [42] or increase in metastatic activity [43].

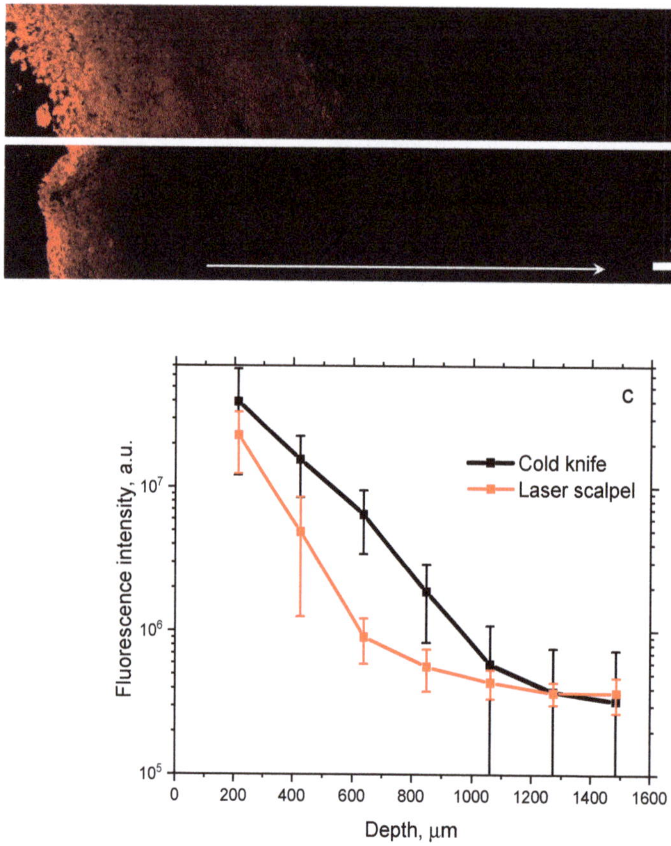

Figure 2. Fluorescence microscopy images of the in-depth PS distribution upon topical application on a tumor after surgical removal of superficial tumor layer with the cold knife (**a**) and the laser scalpel (**b**). White arrow shows in-depth direction. Depth dependence of PS fluorescence upon topical application in muscular tissue within the tumor bed after tumor resection with the cold knife and the laser scalpel (**c**). Graphs show mean values ± standard deviation for 5 animals in each group.

3.3. Monte Carlo Simulations

Another factor that may affect the performance of PDT in deeper tissue layers is the penetration depth of irradiation, which is used to activate the accumulated PS. In this study, with the chlorin-based PS, we compare the effect of PDT performance at the wavelengths of 660 and 405 nm, both of which correspond to chlorin e6 absorption peaks [24] and were proposed for PDT performance [26]. Since the absorbed light dose distribution within tissue cannot be measured directly, we performed Monte Carlo simulations of the absorbed dose distribution for the values of optical properties typical for tumors and muscular tissues summarized in Table 2. Figure 3 shows in-depth profiles of the absorbed dose density calculated for the values typical for tumor (Figure 3a) and muscular (Figure 3b) tissue. The profiles for tumor tissue show faster in-depth attenuation, as compared to muscular tissue owing to the average higher blood content in tumors. The density of the absorbed dose is higher for a wavelength of 405 nm to a depth of about 1 mm for tumors and about 1.6 mm for muscles. Despite strong attenuation at a wavelength of 405 nm, the absorbed doses cannot be neglected until depths of above 1.5 mm and much deeper for a wavelength of 660 nm, indicating that the light of both wavelengths penetrates deeper than PS and can potentially affect deep-seated tumor cells located below.

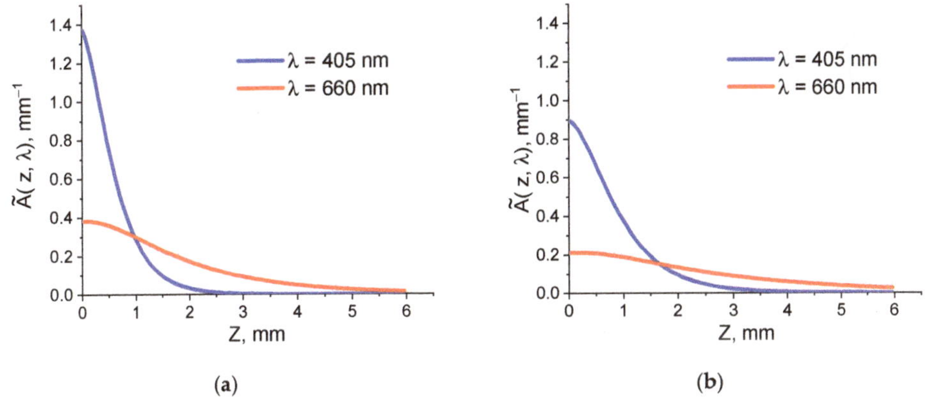

Figure 3. In-depth profiles of the absorbed light dose density for the irradiation wavelengths of 405 and 660 nm for the tumor tissue (**a**) and muscle tissue (**b**) models.

These preliminary studies allow us to conclude that, although PS penetrates quite deep into tissue upon topical application, the caramelization of the superficial tissue layer as a result of the laser scalpel use may prevent deeper PS penetration, thus limiting the impact depth of the PDT procedure. On the other hand, the light at both wavelengths penetrates deeper for both considered models of tumors and muscles. Although these models do not account for the caramelization layer, we do not expect it to significantly affect the optical properties, so these estimates could be treated as reliable for a qualitative analysis.

3.4. Animal Study

Typical photographs of laboratory animals prior to the treatment and in the follow-up period for different treatment protocols are shown in the Supplementary Materials (Figure S1). The results of the treatment outcomes in the considered 10 groups of experimental animals (see Table 1) are summarized in Figure 4. The selected design of the tumor model yielded a zero recurrence rate in the laser scalpel group and 92% recurrence in the cold knife group without post-processing, confirming the efficiency of the laser scalpel in oncology against the cold knife. This result also confirms tumor cell invasion into underlying muscular tissues left after the resection procedure. The application of PDT after the cold knife resection decreased the recurrence rate to 70% and 42% for 405 nm and 660 nm

PDT procedures, respectively. This fact allows one to assume that either the accumulated concentration of PS was insufficient to provide the desired effect or the penetration depth of PS was inadequate for a full response. The fact of much more efficient suppression of recurrence with PDT at 660 nm indicates that light penetration may play a major role in providing the effects of a PDT procedure, so a deeper PDT impact seems to be the key for a successive prevention of recurrence. The observed effect of the recurrence rate decrease is in line with the results with the same tumor line reported in [6], which demonstrated a recurrence rate of 17–33% after PDT of the tumor bed against 83–100% recurrence without PDT. It is worth noting that the study of [6] was performed with another type of PS administered intravenously, which supposedly resulted in a better prevention of recurrence, as compared to our study. Moreover, in that study, wavelengths of 630 and 510 nm were employed for the PDT procedure, providing deeper action, as compared to a wavelength of 405 nm, which demonstrated a weak suppression of recurrence in our study.

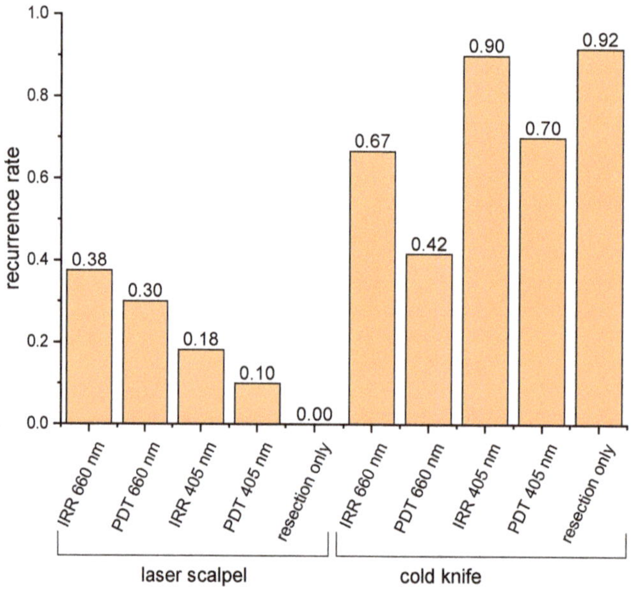

Figure 4. Tumor recurrence rate after surgical removal with the laser scalpel and the cold knife and different regimes of the post-operational PDT of tumor bed (PDT—PDT procedure after topical application of Radagel, IRR—irradiation-only procedure without PS administration).

Surprisingly, the application of PDT after the laser scalpel resection induced recurrence rates of 18% and 30%, respectively, for wavelengths of 405 and 660 nm at the background of total absence of recurrences in the laser-scalpel-only group. This observation allows for assuming that, owing to the limited penetration of PS governed by the presence of the caramelization layer with comparable light penetration, the stimulation of deeper tumor cells may occur. This assumption is confirmed by the observations in the irradiation-only control groups, when post-operation treatment consisted of light irradiation without PS application. In all four irradiation-only groups, the recurrence rate is higher than that in the corresponding PDT groups, indirectly confirming the assumption regarding the stimulation of tumor growth by light in the absence of PS.

However, one can see that the recurrence rate in the irradiation-only group at 405 nm is comparable with the control group, while for the irradiation-only group at 660 nm, it is considerably smaller, which allows us to conclude on the competing processes of suppression and stimulation of tumor recurrence. Supposedly, laser light may suppress recurrence in upper tissue layers, where the fluence rate is rather high, while the low fluence rate may stimulate further tumor development at depth. The potential of the laser light in the partial suppression of tumor growth is also indirectly confirmed by our earlier studies [26], which indicated a delay in tumor growth after irradiation at the employed wavelengths.

On the other hand, previous studies showed that a low-level light illumination may stimulate the proliferation of osteosarcoma, lung carcinoma cells [44], and oral carcinoma cells [45]. This treatment may also modulate inflammation by attenuating TNF-α/cycloheximide-induced apoptosis with a reduction in caspase-3/7/8/9 [46]. It is also worth noting that for the laser scalpel groups, where PS penetration is weaker, the wavelength of 660 nm characterized by deeper penetration provides a higher recurrence rate, as compared to the wavelength of 405 nm, for which the penetration is limited. The cold knife groups demonstrate the opposite effect at the background of deeper PS penetration, where insufficient penetration of irradiation at a wavelength of 405 nm seems to be more critical in providing recurrence rates for the irradiation-only group at 405 nm, comparable with that for the cold knife control group. One can also suppose that the formation of the caramelization layer eliminated the remaining tumor cells in this layer, and recurrence may occur in the tissue below this layer. In this connection, red light, which penetrates deeper, has a higher potential for recurrence stimulation, as compared to blue light at the background of light attenuation in the caramelization layer that results in a low-dose light impact of the underlying tissue.

4. Conclusions

In this paper, we analyzed the efficiency of PDT of the tumor bed as a tool to prevent tumor recurrence after tumor surgical removal. We compared both cold knife and laser scalpel resection, while the choice of the topically applied chlorin-based photosensitizer allowed us to study PDT with red or blue light, which differ significantly in penetration depth in tissue. The initial idea for this study was to provide an additional impact to tumor cells that could remain in tissues after tumor removal. Preliminary fluorescence microscopy studies indicated a significant difference in the distribution of PS in tissues upon topical application for the cold knife and laser scalpel surgery, supposedly owing to the formation of a caramelization layer after the laser surgery. This difference is manifested by a much deeper penetration of PS after the cold knife surgery. Preliminary analysis of the light penetration depth at the two considered wavelengths performed using numerical Monte Carlo simulations indicated deeper penetration of light, as compared to PS penetration in tissues, which allowed us to assume that light could impact deep-seated tumor cells that were not reached by PS. The results of the experiment confirmed some assumptions regarding the effect of PDT of the tumor bed on the tumor recurrence rate; however, some of the revealed effects are surprising. As expected, PDT provided a decrease in the tumor recurrence rate after the cold knife surgery, and the red-light PDT provided a better effect, as compared to the blue-light PDT, owing to the deeper penetration of red light. The irradiation-only groups demonstrated a higher recurrence rate for both red and blue light, additionally confirming the positive effect of PDT.

However, PDT after laser scalpel treatment demonstrated the opposite effect; in contrast to the absence of recurrences in the laser-surgery-only group, both the red- and blue-light PDT after laser scalpel resection demonstrated a certain level of recurrence. This is presumably associated with light stimulation of the remaining tumor cells in the background of weaker penetration of PS through the caramelization layer upon topical application. Higher recurrence rates in the irradiation-only groups after the laser scalpel resection as compared to the corresponding PDT groups confirm the stimulating impact of irradiation in the absence of PS. Thus, one should be careful when deciding to improve the surgical tumor removal with post-operational PDT. It seems that a sufficient delivery of PS to the remaining tumor cells should be the key element of this procedure. Perhaps some additional techniques should be employed to confirm sufficient accumulation of PS in tissue depth; alternatively, a traditional approach of intravenous PS injection should be considered.

Supplementary Materials: The following supporting information can be downloaded at: https://www.mdpi.com/article/10.3390/biomedicines12020291/s1, Figure S1: Photographs of laboratory animals before tumor resection (upper row) and in the follow up (lower row). Animals with recurrences (IRR 660 nm for laser scalpel and all examples for cold knife) were documented 3 weeks after resection. Animals without recurrences were documented 3 months after resection (examples for laser scalpel, except IRR 660 nm). The treatment protocols are indicated under each column (IRR = irradiation only, PDT = PDT procedure, irradiation wavelength is shown next to the treatment option, the irradiation dose was 150 J/cm^2).

Author Contributions: Conceptualization, V.K. and M.K.; methodology, M.S. and V.E.; software, D.K.; validation, A.K.; formal analysis, A.P.; investigation, M.S., V.E., A.P. and A.K.; writing—original draft preparation, M.K., V.E., A.P. and V.K.; writing—review and editing, M.K.; supervision, V.K.; project administration, D.K.; funding acquisition, V.K. All authors have read and agreed to the published version of the manuscript.

Funding: This research was supported by the Center of Excellence "Center of Photonics" funded by the Ministry of Science and Higher Education of the Russian Federation, contract no. 075-15-2022-316.

Institutional Review Board Statement: The animals were kept in accordance with the rules adopted by the European convention for the protection of vertebrate animals used for experimental and other scientific purposes. The experimental study was approved by the Ethics committee of the Privolzhsky Research Medical University (Protocol No. 13 of 7 July 2021).

Informed Consent Statement: Not applicable.

Data Availability Statement: The data presented in this study are available on request from the corresponding author.

Acknowledgments: The authors are thankful to Liubov Shimolina (Privolzhsky Research Medical University) for her help in the animal monitoring.

Conflicts of Interest: The authors declare no conflicts of interest.

References

1. Shao, N.; Wei, X.; Zhang, Y.; Luo, H.; Su, Y.; Liang, L.; Chen, B.; Li, N.; Ren, X.; Yang, H. Effect of different surgical modalities on swallowing-related quality of life in patients with glottic laryngeal squamous cell carcinoma: How should we choose? *Arch. Med. Sci.* **2023**, *19*, 550–554. [CrossRef]
2. Hans, S.; Baudouin, R.; Circiu, M.P.; Couineau, F.; Lisan, Q.; Crevier-Buchman, L.; Lechien, J.R. Laryngeal Cancer Surgery: History and Current Indications of Transoral Laser Microsurgery and Transoral Robotic Surgery. *J. Clin. Med.* **2022**, *11*, 5769. [CrossRef]
3. Amaral, M.B.; de Avila, J.M.; Abreu, M.H.; Mesquita, R.A. Diode laser surgery versus scalpel surgery in the treatment of fibrous hyperplasia: A randomized clinical trial. *Int. J. Oral. Maxillofac. Surg.* **2015**, *44*, 1383–1389. [CrossRef]
4. Jia, W.; King, E. The Role of Robotic Surgery in Laryngeal Cancer. *Otolaryngol. Clin. N. Am.* **2023**, *56*, 313–322. [CrossRef]
5. Mahvi, D.A.; Liu, R.; Grinstaff, M.W.; Colson, Y.L.; Raut, C.P. Local Cancer Recurrence: The Realities, Challenges, and Opportunities for New Therapies. *CA Cancer J. Clin.* **2018**, *68*, 488–505. [CrossRef]
6. Abulafi, A.M.; Dejode, M.; Allardice, J.T.; Ansell, J.; Rogers, J.; Williams, N.S. Adjuvant intraoperative photodynamic therapy in experimental colorectal cancer. *Br. J. Surg.* **1995**, *82*, 178–181. [CrossRef]
7. Nanashima, A.; Yamaguchi, H.; Shibasaki, S.; Ide, N.; Sawai, T.; Tsuji, T.; Hidaka, S.; Sumida, Y.; Nakagoe, T.; Nagayasu, T. Adjuvant photodynamic therapy for bile duct carcinoma after surgery: A preliminary study. *J. Gastroenterol.* **2004**, *39*, 1095–1101. [CrossRef]
8. Agostinis, P.; Berg, K.; Cengel, K.A.; Foster, T.H.; Girotti, A.W.; Gollnick, S.O.; Hahn, S.M.; Hamblin, M.R.; Juzeniene, A.; Kessel, D.; et al. Photodynamic therapy of cancer: An update. *CA Cancer J. Clin.* **2011**, *61*, 250–281. [CrossRef]
9. Fang, L.; Zhao, Z.; Wang, J.; Zhang, P.; Ding, Y.; Jiang, Y.; Wang, D.; Li, Y. Engineering autologous tumor cell vaccine to locally mobilize antitumor immunity in tumor surgical bed. *Sci. Adv.* **2020**, *6*, eaba4024. [CrossRef]
10. Lamy, L.; Thomas, J.; Leroux, A.; Bisson, J.-F.; Myren, K.; Godal, A.; Stensrud, G.; Bezdetnaya, L. Antitumor effect and induced immune response following expo-sure of hexaminolevulinate and blue light in combination with checkpoint inhibitor in an orthotopic model of rat bladder cancer. *Biomedicines* **2022**, *10*, 548. [CrossRef]
11. Giuliano, E.A.; Johnson, P.J.; Delgado, C.; Pearce, J.W.; Moore, C.P. Local photodynamic therapy delays recurrence of equine periocular squamous cell carcinoma compared to cryotherapy. *Vet. Ophthalmol.* **2014**, *17* (Suppl. S1), 37–45. [CrossRef] [PubMed]

12. Luo, X.M.; Niu, L.Z.; Chen, J.B.; Xu, K.C. Advances in cryoablation for pancreatic cancer. *World J. Gastroenterol.* **2016**, *22*, 790–800. [CrossRef] [PubMed]
13. Mokbel, K.; Kodresko, A.; Ghazal, H.; Mokbel, R.; Trembley, J.; Jouhara, H. The Evolving Role of Cryosurgery in Breast Cancer Management: A Comprehensive Review. *Cancers* **2023**, *15*, 4272. [CrossRef] [PubMed]
14. Osintsev, A.M.; Vasilchenko, I.L.; Rodrigues, D.B.; Stauffer, P.R.; Braginsky, V.I.; Rynk, V.V.; Gromov, E.S.; Prosekov, A.Y.; Kaprin, A.D.; Kostin, A.A. Characterization of Ferromagnetic Composite Implants for Tumor Bed Hyperthermia. *IEEE Trans. Magn.* **2021**, *57*, 5400108. [CrossRef] [PubMed]
15. Cheng, Y.; Weng, S.; Yu, L.; Zhu, N.; Yang, M.; Yuan, Y. The Role of Hyperthermia in the Multidisciplinary Treatment of Malignant Tumors. *Integr. Cancer Ther.* **2019**, *18*, 1534735419876345. [CrossRef] [PubMed]
16. Fassnacht, M.; Hahner, S.; Polat, B.; Koschker, A.-C.; Kenn, W.; Flentje, M.; Allolio, B. Efficacy of adjuvant radiotherapy of the tumor bed on local recurrence of adrenocortical carcinoma. *J. Clin. Endocrinol. Metab.* **2006**, *91*, 4501–4504. [CrossRef]
17. Haas, R.L.; Walraven, I.; Lecointe-Artzner, E.; van Houdt, W.J.; Strauss, D.; Schrage, Y.; Hayes, A.J.; Raut, C.P.; Fairweather, M.; Baldini, E.H.; et al. Extrameningeal solitary fibrous tumors—Surgery alone or surgery plus perioperative radiotherapy: A retrospective study from the global solitary fibrous tumor initiative in collaboration with the Sarcoma Patients EuroNet. *Cancer* **2020**, *126*, 3002–3012. [CrossRef]
18. Guan, X.; Sun, L.; Shen, Y.; Jin, F.; Bo, X.; Zhu, C.; Han, X.; Li, X.; Chen, Y.; Xu, H.; et al. Nanoparticle-enhanced radiotherapy synergizes with PD-L1 blockade to limit post-surgical cancer recurrence and metastasis. *Nat. Commun.* **2022**, *13*, 2834. [CrossRef]
19. Tong, C.W.; Wu, W.K.; Loong, H.H.; Cho, W.C.; To, K.K. Drug combination approach to overcome resistance to EGFR tyrosine kinase inhibitors in lung cancer. *Cancer Lett.* **2017**, *405*, 100–110. [CrossRef]
20. Dolmans, D.E.; Fukumura, D.; Jain, R.K. Photodynamic therapy for cancer. *Nat. Rev. Cancer* **2003**, *3*, 380–387. [CrossRef]
21. Park, J.; Lee, Y.K.; Park, I.K.; Hwang, S.R. Current Limitations and Recent Progress in Nanomedicine for Clinically Available Photodynamic Therapy. *Biomedicines* **2021**, *9*, 85. [CrossRef]
22. Momma, T.; Hamblin, M.R.; Wu, H.C.; Hasan, T. Photodynamic therapy of orthotopic prostate cancer with benzoporphyrin derivative: Local control and distant metastasis. *Cancer Res.* **1998**, *58*, 5425–5431.
23. Grandi, V.; Sessa, M.; Pisano, L.; Rossi, R.; Galvan, A.; Gattai, R.; Mori, M.; Tiradritti, L.; Bacci, S.; Zuccati, G.; et al. Photodynamic therapy with topical photosensitizers in mucosal and semimucosal areas: Review from a dermatologic perspective. *Photodiagnosis Photodyn. Ther.* **2018**, *23*, 119–131. [CrossRef]
24. Kirillin, M.; Khilov, A.; Kurakina, D.; Orlova, A.; Perekatova, V.; Shishkova, V.; Malygina, A.; Mironycheva, A.; Shlivko, I.; Gamayunov, S.; et al. Dual-Wavelength Fluorescence Monitoring of Photodynamic Therapy: From Analytical Models to Clinical Studies. *Cancers* **2021**, *13*, 5807. [CrossRef]
25. Shakhova, M.; Loginova, D.; Meller, A.; Sapunov, D.; Orlinskaya, N.; Shakhov, A.; Khilov, A.; Kirillin, M. Photodynamic therapy with chlorin-based photosensitizer at 405 nm: Numerical, morphological, and clinical study. *J. Biomed. Opt.* **2018**, *23*, 091412. [CrossRef]
26. Kirillin, M.; Kurakina, D.; Khilov, A.; Orlova, A.; Shakhova, M.; Orlinskaya, N.; Sergeeva, E. Red and blue light in antitumor photodynamic therapy with chlorin-based photosensitizers: A comparative animal study assisted by optical imaging modalities. *Biomed. Opt. Express.* **2021**, *12*, 872–892. [CrossRef]
27. Zeitouni, N.C.; Bhatia, N.; Ceilley, R.I.; Cohen, J.L.; Del Rosso, J.Q.; Moore, A.Y.; Munavalli, G.; Pariser, D.M.; Schlesinger, T.; Siegel, D.M.; et al. Photodynamic Therapy with 5-aminolevulinic Acid 10% Gel and Red Light for the Treatment of Actinic Keratosis, Nonmelanoma Skin Cancers, and Acne: Current Evidence and Best Practices. *J. Clin. Aesthet. Dermatol.* **2021**, *14*, E53–E65.
28. Yang, H.; Zhang, Z.; Zhao, K.; Zhang, Y.; Yin, X.; Zhu, G.; Wang, Z.; Sui, Y.; Li, X.; Li, C.; et al. Initial Experience with Extraperitoneal Laparoscopic Radical Cystectomy With Pelvic Organ-Preserving and Orthotopic Neobladder Techniques for Bladder Cancer in Female Patients. *Urology* **2023**, *171*, 77–82. [CrossRef]
29. Hanke, A.; Fimmers, R.; Frentzen, M.; Meister, J. Quantitative determination of cut efficiency during soft tissue surgery using diode lasers in the wavelength range between 400 and 1500 nm. *Lasers Med. Sci.* **2021**, *36*, 1633–1647. [CrossRef]
30. Sapogova, N.; Bredikhin, V.; Bityurin, N.; Kamensky, V.; Zhigarcov, V.; Yusupov, V. Model for indirect laser surgery. *Biomed. Opt. Express* **2016**, *8*, 104–111. [CrossRef]
31. Kuznetsova, D.S.; Karabut, M.M.; Elagin, V.V.; Shakhova, M.A.; Bredikhin, V.I.; Baskina, O.S.; Snopova, L.B.; Shakhov, A.V.; Kamensky, V.A. Comparative Analysis of Biotissue Laser Resection Using Strongly Absorbing Optical Fiber Tips. *Opt. Photonics J.* **2015**, *5*, 1–5. [CrossRef]
32. Khilov, A.; Kirillin, M.; Loginova, D.; Turchin, I. Estimation of chlorin-based photosensitizer penetration depth prior to photodynamic therapy procedure with dual-wavelength fluorescence imaging. *Laser Phys. Lett.* **2018**, *15*, 126202. [CrossRef]
33. Greening, G.; Mundo, A.; Rajaram, N.; Muldoon, T.J. Sampling depth of a diffuse reflectance spectroscopy probe for in-vivo physiological quantification of murine subcutaneous tumor allografts. *J. Biomed. Opt.* **2018**, *23*, 085006. [CrossRef]
34. Honda, N.; Ishii, K.; Terada, T.; Nanjo, T.; Awazu, K. Determination of the tumor tissue optical properties during and after photodynamic therapy using inverse Monte Carlo method and double integrating sphere between 350 and 1000 nm. *J. Biomed. Opt.* **2011**, *16*, 058003. [CrossRef]
35. Loginova, D.A.; Sergeeva, E.A.; Krainov, A.D.; Agrba, P.D.; Kirillin, M.Y. Liquid optical phantoms mimicking spectral characteristics of laboratory mouse biotissues. *Quantum Electron.* **2016**, *46*, 528. [CrossRef]

36. Zabotnov, S.V.; Skobelkina, A.V.; Sergeeva, E.A.; Kurakina, D.A.; Khilov, A.V.; Kashaev, F.V.; Kaminskaya, T.P.; Presnov, D.E.; Agrba, P.D.; Shuleiko, D.V.; et al. Nanoparticles Produced via Laser Ablation of Porous Silicon and Silicon Nanowires for Optical Bioimaging. *Sensors* **2020**, *20*, 4874. [CrossRef]
37. Plekhanov, A.A.; Sirotkina, M.A.; Sovetsky, A.A.; Gubarkova, E.V.; Kuznetsov, S.S.; Matveyev, A.L.; Matveev, L.A.; Zagaynova, E.V.; Gladkova, N.D.; Zaitsev, V.Y. Histological validation of in vivo assessment of cancer tissue inhomogeneity and automated morphological segmentation enabled by Optical Coherence Elastography. *Sci. Rep.* **2020**, *10*, 11781. [CrossRef]
38. Elagin, V.V.; Shakhova, M.A.; Sirotkina, M.A.; Shakhov, A.V.; Pavlova, N.P.; Snopova, L.B.; Bredikhin, V.I.; Kamensky, V.A. Can "Indirect" Contact Laser Surgery be Used for Fluorescence-Image Guided Tumor Resections? Preliminary Results. *Technol. Cancer Res. Treat.* **2018**, *17*, 1533033818805715. [CrossRef]
39. Berger, N.A.; Eeg, P.H. *Veterinary Laser Surgery: A Practical Guide*; Wiley-Blackwell: Hoboken, NJ, USA, 2006; 256p.
40. Svaasand, L.O.; Wyss, P.; Wyss, M.-T.; Tadir, Y.; Tromberg, B.J.; Berns, M.W. Dosimetry model for photodynamic therapy with topically administered photosensitizers. *Lasers Surg. Med.* **1996**, *18*, 139–149. [CrossRef]
41. Li, L.-B.; Xie, J.-M.; Zhang, X.-N.; Chen, J.-Z.; Luo, Y.-L.; Zhang, L.-Y.; Luo, R.-C. Retrospective study of photodynamic therapy vs photodynamic therapy combined with chemotherapy and chemotherapy alone on advanced esophageal cancer. *Photodiagnosis Photodyn. Ther.* **2010**, *7*, 139–143. [CrossRef]
42. Pech, O.; Behrens, A.; May, A.; Nachbar, L.; Gossner, L.; Rabenstein, T.; Manner, H.; Guenter, E.; Huijsmans, J.; Vieth, M.; et al. Long-term results and risk factor analysis for recurrence after curative endoscopic therapy in 349 patients with high-grade intraepithelial neoplasia and mucosal adenocarcinoma in Barrett's oesophagus. *Gut* **2008**, *57*, 1200–1206. [CrossRef]
43. Yujie, S.; Mingming, L.; Fang, S.; Zhang, Y.; Qu, C.; Zhou, M.; Shen, F.; Xu, L. Low-dose photodynamic therapy-induced increase in the metastatic potential of pancreatic tumor cells and its blockade by simvastatin. *J. Photochem. Photobiol. B Biol.* **2020**, *207*, 111889.
44. Kara, C.; Selamet, H.; Gökmenoğlu, C.; Kara, N. Low level laser therapy induces increased viability and proliferation in isolated cancer cells. *Cell Prolif.* **2018**, *51*, e12417. [CrossRef]
45. De Castro, J.L.; Pinheiro, A.L.; Werneck, C.E.; Soares, C.P. The effect of laser therapy on the proliferation of oral KB carcinoma cells: An in vitro study. *Photomed. Laser Surg.* **2005**, *23*, 586–589. [CrossRef]
46. Chu, Y.-H.; Chen, S.-Y.; Hsieh, Y.-L.; Teng, Y.-H.; Cheng, Y.-J. Low-level laser therapy prevents endothelial cells from TNF-α/cycloheximide-induced apoptosis. *Lasers Med. Sci.* **2018**, *33*, 279–286. [CrossRef]

biomedicines

MDPI

Review

Current Photodynamic Therapy for Glioma Treatment: An Update

David Aebisher [1,*], Agnieszka Przygórzewska [2], Angelika Myśliwiec [3], Klaudia Dynarowicz [3], Magdalena Krupka-Olek [4], Andrzej Bożek [4], Aleksandra Kawczyk-Krupka [5,*] and Dorota Bartusik-Aebisher [6]

1 Department of Photomedicine and Physical Chemistry, Medical College of the Rzeszów University, 35-959 Rzeszów, Poland
2 English Division Science Club, Medical College of the Rzeszów University, 35-025 Rzeszów, Poland; ap117623@stud.ur.edu.pl
3 Center for Innovative Research in Medical and Natural Sciences, Medical College of the Rzeszów University, 35-310 Rzeszów, Poland; amysliwiec@ur.edu.pl (A.M.); kdynarowicz@ur.edu.pl (K.D.)
4 Clinical Department of Internal Medicine, Dermatology and Allergology, Medical University of Silesia in Katowice, M. Sklodowskiej-Curie 10, 41-800 Zabrze, Poland; magda.krupka94@gmail.com (M.K.-O.); andrzej.bozek@sum.edu.pl (A.B.)
5 Department of Internal Medicine, Angiology and Physical Medicine, Center for Laser Diagnostics and Therapy, Medical University of Silesia in Katowice, Batorego 15 Street, 41-902 Bytom, Poland
6 Department of Biochemistry and General Chemistry, Medical College of the Rzeszów University, 35-025 Rzeszów, Poland; dbartusikaebisher@ur.edu.pl
* Correspondence: daebisher@ur.edu.pl (D.A.); akawczyk@sum.edu.pl (A.K.-K.)

Abstract: Research on the development of photodynamic therapy for the treatment of brain tumors has shown promise in the treatment of this highly aggressive form of brain cancer. Analysis of both in vivo studies and clinical studies shows that photodynamic therapy can provide significant benefits, such as an improved median rate of survival. The use of photodynamic therapy is characterized by relatively few side effects, which is a significant advantage compared to conventional treatment methods such as often-used brain tumor surgery, advanced radiotherapy, and classic chemotherapy. Continued research in this area could bring significant advances, influencing future standards of treatment for this difficult and deadly disease.

Keywords: glioma; PDT; cancer; treatment

1. Introduction

Gliomas are internal brain tumors originating from neuroglial progenitor cells [1,2]. They are highly heterogeneous tumors that are resistant to treatment [3,4]. Gliomas account for approximately 80% of the malignant intracranial tumors diagnosed [5–8]. Although gliomas are relatively rare in the population, they are associated with significant mortality with a 5-year relative survival rate of approximately 5% [9–12]. Gliomas invade and attack brain parenchyma, white matter tracts, and perivascular spaces [13]. The glioma tumor microenvironment (TME) consists of stromal cells, tumor components, and innate and adaptive immune cells, and these cells, along with the extracellular matrix, regulate and communicate intracellularly to promote TME formation. The immune microenvironment plays an important role in the development of glioma [14]. Patients with glioma are usually diagnosed at the symptomatic stage of the disease [15]. Although there is currently no cure for malignant glioma, researchers around the world continue to pursue a deeper understanding of the factors that influence glioma development and response to treatment [16]. Despite advances in neurosurgical technology and techniques, the survival rate of glioma patients has remained relatively unchanged in recent years, and thus improving the efficacy of glioma treatment is an urgent task in medicine. The main treatments for gliomas include surgery, radiotherapy, and chemotherapy [17–21]. Unfortunately, these methods are often associated with difficulties resulting from incomplete tumor resection and recurrence [22–24]. In such situations, photodynamic therapy (PDT) is becoming increasingly popular as an advanced therapeutic strategy, characterized by fewer side effects, minimal toxicity, and more controlled treatment [24–26]. The first studies on the use of PDT in high-grade gliomas showed promising results, such as increasing the median survival and extending the 2-year survival of patients from 18% to 28% [27–31]. In the initial phase of the development of PDT in oncology, the main goal was the complete removal of localized tumors, but over time, the clinical application of PDT in cancer treatment began to change [32–36]. Photodynamic therapy can be combined with additional anticancer therapies used in the clinic [37–43]. Successful treatment of glioma is hampered by many obstacles, including the immunosuppressive tumor microenvironment, the blood–brain barrier, and high heterogeneity [44]. Therefore, overcoming sufficient drug delivery and drug targeting to the glioma area is the key to obtaining an efficient treatment [45]. Continuing research on this innovative approach in the fight against malignant gliomas is certainly a step in the right direction and may result in further important discoveries. Novelty in our review refers to the unique perspective regarding the application of PDT in clinics. Photodynamic therapy addresses some hindrances in glioma treatment, e.g., the heterogeneous nature of treatment sites, and is in need of improvements in light delivery, imaging methods for determining PDT efficiency, and treatment planning.

2. Methodology

Publications written in English on the topic of PDT for glioma were searched for in PubMed and Research Gate. The keywords used for the search were: glioma; PDT; cancer; and treatment. Our article selection was based on the following criteria: (1) recent major achievements in this field; (2) current debates; and (3) ideas for future development of PDT for glioma. The papers not selected for this review included (1) technical reports, (2) papers not written in English, (3) clinical cases, and (4) PDT performed in vitro.

3. Standard Strategies for the Treatment of Gliomas

The greatest successes in the treatment of brain tumors are achieved by centers that implement multifaceted treatments of these types of tumors with participation of specialists in neurosurgery, radiology, pathology, radiation oncology, and neuro-oncology. There are three established guidelines for glioma treatment: surgery, radiotherapy, chemotherapy, and targeted therapy [46,47].

3.1. Surgery

The use of surgery in eradicating glioma faces problems of complete resection of glioma. These current problems are associated with imaging techniques to differentiate glioma and healthy brain tissue, resection extent, neurological function preservation, and tumor margin evaluation [48,49].

Despite inevitable recurrences, tumor resection is still an important element in prolonging patients' lives. Technological advances that enable real-time visualization and quantification of tumorous and non-tumorous tissue are critical for brain tumor surgery [50]. Therefore, surgical planning by carefully observing the relationship between the area of glioma invasion and the eloquent area of connecting fibers is crucial. Neurosurgeons typically detect the eloquent area using functional MRI and identify the connecting fiber using diffusion tensor imaging [51]. Studies suggest that more extensive surgical resection may be associated with improved life expectancy in both low-grade and high-grade glioma patients [52,53]. Treatment strategies for patients with intracerebral glioma must be patient-specific. Studies indicate that maximal resection is desirable in patients with low-grade gliomas while preserving neurologic function. In patients with high-grade gliomas, the benefits of aggressive resection are more difficult to confirm. Patients who are relatively young and have tumors of significant mass are those who benefit most from aggressive resection [54]. In the recent neurosurgical literature, neurological outcomes were most frequently reported, followed by activities of daily living, seizure outcomes, neurocognitive outcomes, and Health-Related Quality of Life (HRQoL) outcomes following glioma surgery [55].

3.2. *Radiotherapy*

The use of conventional radiotherapy (RT) is the mainstay for the treatment of glioma. Although this methodology is advanced, it is also faced with problems. Despite the use of radiotherapy, chemotherapy, and surgical resection, glioma retains a poor prognostic value and no significant improvement in survival rates has been observed over the last 20 years [56].

Fractionated local irradiation is the main way to treat glioma. A standard radiation dose of 60 Gy is administered in more than 30 fractions of 1.8–2 Gy. The therapy is targeted at the diseased area with a 1–3 cm margin to effectively treat the infiltrating tumor [52]. Clinical RT was designed to target the majority of remaining glioma cells in healthy brain tissue with fewer side effects [57]. It is believed that the post-irradiation microenvironment is not suitable for the survival of cancer cells (tumor bed effect). However, resistance to radiotherapy remains a serious problem [58]. New techniques for delivering radiotherapy to the tumor are currently under investigation [59].

3.3. *Chemotherapy*

Chemotherapy is used before surgery in order to shrink the tumor prior to or after surgery to destroy the remaining cancer cells. To destroy glioma, high doses of chemo drugs are administered and, as a consequence, many healthy cells are also destroyed; stem cells can be transplanted into the patient [60].

Currently, temozolomide and chloroethyl nitroureas are mainly used to treat glioma with chemotherapy. Unfortunately, drugs administered systemically do not usually achieve high concentrations in the central nervous system and tumor area, leading to significant systemic side effects. Side effects of chemotherapy on the central nervous system include acute encephalopathy, leukoencephalopathy, cerebellar dysfunction, and spinal cord toxicity. Additionally, side effects may also occur in the peripheral nervous system, manifesting as chemotherapy-induced peripheral neuropathic pain [61,62]. Temozolomide (TMZ) is a common alkylating chemotherapeutic agent used to treat brain tumors such as glioma and anaplastic astrocytoma [63]. Temozolomide has a small molecular mass and this alkylating derivative directly damages brain tumor cells by methylating DNA [64]. The key cytotoxic effect is the formation of O^6-methylguanine, which causes apoptosis, autophagy, and cell senescence [65,66]. Although TMZ therapy has shown gradual improvement in the treatment of high-grade gliomas, for most patients it is mainly a palliative treatment. In fact, in newly diagnosed glioma patients, the median survival benefit with TMZ plus radiation therapy is only 2.5 months compared with radiation therapy alone. Recent studies also suggest that 60–75% of glioma patients derive no benefit from TMZ treatment [67]. Most patients do not respond to TMZ during treatment. Activation of DNA repair pathways is the main mechanism of this phenomenon, which disconnects TMZ-induced O^6-methylguanine adducts and restores genome integrity. Current knowledge in the field of oncology adds several other new resistance mechanisms, such as the involvement of miRNAs, drug efflux transporters, gap junction activity, the emergence of glioma stem cells, and the enhancement of cell survival autophagy [68]. Chloroethyl nitroureas (CNU) are DNA alkylating agents. In clinical practice, the most commonly used CNUs are lomustine, carmustine, and fotemustine [69].

In patients with WHO grade 4 gliomas and astrocytomas who have relapsed, lomustine chemotherapy is the standard treatment [70]. Lomustine, also known as CCNU, is an alkylating agent from the nitrosourea family [71,72]. The most common side effects are thrombocytopenia, with neutropenia and lymphocytopenia occurring less frequently [73–75]. Carmustine is used both in the initial diagnosis of glioma and in the case of tumor recurrence, both by intravenous administration [76]. The use in clinical practice is lower due to particularly long-lasting bone marrow suppression and persistent lung toxicity [77,78]. The developed biodegradable carmustine wafer may prove significant, enabling the administration of high doses of drugs [79,80]. Fotemustine has been used in the treatment of melanoma and is currently being tested for its effectiveness in the treatment of recurrent malignancy glioma [81].

3.4. Targeted Treatment

Tumor targeting as a novel strategy for glioma and targeted therapy uses drugs or other substances to identify and destroy cancer cells without damaging normal healthy cell survival and overall survival [82].

Pathways involved in tumor growth and processes of invasion and angiogenesis represent the main processes in glioma. The goal is to individualize treatment based on the specific molecular abnormalities of a given tumor [52]. Most drugs that target growth and survival pathways as monotherapies have failed to demonstrate survival benefits in populations of glioma patients. Targeting multiple signaling pathways with multi-target kinase inhibitors or combinations of single-target kinase inhibitors may increase treatment efficacy [83]. Main targets, including p53, retinoblastoma, and epidermal growth factor receptor gene, have been demonstrated to be ineffective. Research continues in immunotherapy-based treatments using immune checkpoint molecules, macrophages, and dendritic cells within the tumor microenvironment [84]. In recent years, CTLA-4 inhibitors have shown good effects in cancer immunotherapy. Tremelimumab is a monoclonal antibody against CTLA4 that has demonstrated positive responses in many published clinical trials when combined with the PD-1/PD-L1 blockade. However, in glioma, several studies have shown that antibodies anti-CTLA-4 and PD-1 show no survival benefit when compared to standard chemotherapy [85–88]. Furthermore, no obvious benefit was obtained with neoadjuvant nivolumab in resectable glioma [89]. Similarly, a phase III trial comparing nivolumab (anti-PD-1 blocking antibody) with bevacizumab (anti-VEGF blocking antibody) in patients with recurrent glioma showed no benefit from nivolumab and resulted in a similar median overall survival (mOS 9.8 vs. 10.0 months) [90,91]. Nevertheless, the paucity of effective antigenic targets remains a significant obstacle to the safe and effective treatment of gliomas with relatively low mutational load with immunotherapy [92]. Merely reducing the size of the tumor is not sufficient because recurrence and rapid progression of the tumor will ultimately kill the patient. There is a trend to incorporate immunotherapy into multimodal treatments, including radiotherapy and chemotherapy in patients with glioma because the effects of individual treatments may enhance each other [93]. Bevacizumab treatment is administered intravenously and this treatment with bevacizumab appears to not significantly improve overall survival in patients with newly diagnosed glioma [66,94]. The use of bevacizumab with irinotecan is also being investigated [95]. The most common side effects of bevacizumab include hypertension and leukopenia [96,97]. There is no evidence of differences in the effectiveness of chemotherapy depending on the age, sex, histology, functional status, or extent of resection [98].

4. Photodynamic Therapy

4.1. Mechanism of Tumor Destruction Using the PDT Method

Photodynamic therapy employs photosensitizer molecules (PSs) that are selectively introduced into cancer cells. There are two key mechanisms of action in PDT. In the first mechanism (Type I), a photon absorbed by a photosensitizer molecule causes its excitation from the singlet ground state to the singlet excited state (PS*). The photosensitizer then inter-system crosses to the excited triplet state where it can undergo electron transfer reactions with oxygen or biomolecules in the environment. In the second mechanism (Type II), the photosensitizer in the excited triplet state transfers energy directly to the oxygen molecule to produce singlet oxygen ($^1O_2$). This ROS is characterized by a short lifetime and diffusion distance. Singlet oxygen is an electrophile and reacts rapidly with double bonds and heteroatoms such as sulfur. These reactions oxidatively damage the membranes of intracellular organelles such as mitochondria, lysosomes, and the endoplasmic reticulum [99–101]. This can promote irreversible destruction of cancer tissue primarily by apoptosis and/or necrosis (Figure 1). Another mechanism involves anti-vascular effects with the activation of PSs, provoking destruction of tumor vasculature. The third mechanism involves the activation of the immune response against cancer cells through acute inflammatory processes and the release of cytokines into the tumor. The advantage of one pathway over the other depends on the parameters used in therapy and the patient's health condition [102–108].

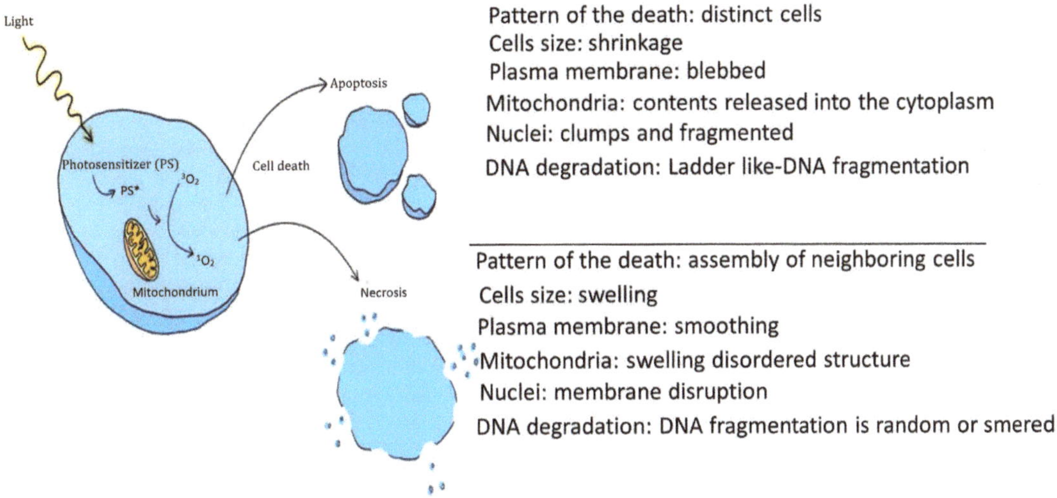

Figure 1. Mechanism of cell death after photodynamic therapy as a result of apoptosis and necrosis.

4.2. PDT in Cancer Treatment

Clinical applications of PDT in dermatology, ophthalmology, urology, gastroenterology, gynecology, neurosurgery, and pneumology have shown encouraging results in the treatment of human cancers [103]. It is less invasive, target-specific, and has reduced cytotoxicity towards normal cells and tissues, which translates into fewer side effects [109–111]. Photodynamic therapy is selective as only diseased tissues that have accumulated PSs are irradiated. The idea of using PDT as a new treatment strategy appeared at the beginning of the 20th century [112–114]. Photodynamic therapy is a two-step procedure. First, a drug that absorbs light is administered to the patient. Then, after a period of time called the drug-light interval, the target tissue is irradiated. The drug is inactive in the dark, but upon electronic excitation, it transfers energy to molecular oxygen in the Type II process described previously [109,115–120]. The molecular structure and chemical properties of a PS determine the required wavelength of light, effective doses, and mechanism of action. The selection of an appropriate PS is one of the most important factors in achieving the intended effect [103,121]. Although the results of PDT vary depending on the type of cancer, its properties make it efficient in therapies [122].

4.3. Photodynamic Therapy in the Treatment of Brain Gliomas

Photodynamic therapy is applied for the treatment of neurological diseases and many types of brain tumors [123]. Photodynamic therapy has fewer side effects compared to chemotherapy and radiotherapy on the brain. After administration, the concentration of PSs is higher in glioma cells than in healthy tissue [31,124]. When combined with surgical resection, PDT for the treatment of gliomas, or alternatively as a stand-alone treatment strategy, has had some success in extending median patient survival compared to surgery alone [125]. The immunological effects of PDT are of particular interest given recent studies demonstrating the importance of these processes in glioma [126,127]. However, the use of this method in the treatment of gliomas also has some limitations. The main drawback is the development of resistance to PDT in tumors. Several mechanisms are known to be involved in the development of cellular defense against the cytotoxic effects of PDT, including activation of antioxidant enzymes, drug efflux pumps, PS degradation, and overexpression of chaperones [128].

DNA repair may aid in glioma resistance to PDT; however, this has not been further explored to date [129]. Many biological barriers may have an influence on the results of glioma PDT [9]. These include technical limitations of light delivery [130]. However, the problem of delivering light to the tumor, at least in some cases, can be solved by using implantable devices that enable light delivery during PDT, or near-infrared lasers that allow tissue penetration of up to 3 cm [9]. The effectiveness of glioma-PDT is based on the activation of PSs accumulated in the tumor with light. However, insufficient accumulation of PSs in the tumor severely limits the success of PDT [131]. The blood–brain barrier (BBB) is a significant limitation of PS transport to the area of postoperative resection, where brain tumor recurrence most often occurs [132,133]. In order to develop the "ideal photosensitizer", there is still a need for new photodynamic agents with improved photophysical and photobiological properties [134]. Recent research has also led to the discovery of profound genetic heterogeneity among glioma cells that includes the adaptation to ROS. Therefore, tumor heterogeneity and the associated difference in sensitivity to ROS-producing therapeutic agents must be taken into account when designing PDT protocols to predict outcomes [135]. Moreover, there are no standard guidelines for PDT treatment protocols, and it is known that the selection of parameters affects the quality of treatment. Further observations are needed to further assess how PDT will reduce morbidity and mortality [136–138].

4.4. Nanoparticles for Glioma and PDT

To improve the results of brain tumor treatment, the use of nanoparticles containing a photosensitizer may be a promising strategy [9,139,140]. They are characterized by low cytotoxicity and excellent light absorption, which make them excellent agents for improving multifunctionality for imaging and treatment [141,142]. Numerous studies are currently underway to discover a nanoparticle with optimal properties for the treatment of glioma. Comincini et al. developed nanoparticles equipped with the photosensitizer berberine and showed that PDT using this system was responsible for an increase in early and late apoptosis of cancer cells without detecting any cytotoxic effect on healthy tissue [143]. Liu et al. constructed glycolipid-like micelles containing indocyanine green and demonstrated that it has the ability to improve drug delivery to neovascular endothelial cells and tumor cells and increase the effectiveness of phototherapy [144]. Teng et al. created nanoparticles by combining Indocyanine-Green with Chlorin-e6 on the surface of superparamagnetic iron oxide nanoparticles (SPIONs) and showed that they significantly increased local control of recurrent cancer and that they could be detected by NIR imaging, creating the potential for their use also in surgical oncology [145].

Research on upconversion nanoparticles (UCN) with covalently attached PSs has been conducted [146–157]. Upconversion nanoparticles are nanotransducers (NT) that absorb near-infrared light (NIR) and emit visible light (fluorescence) at wavelengths absorbed by PSs.

4.5. Photosensitizers Used in PDT

Obtaining new photosensitizers is a promising direction of research on photodynamic therapy (PDT) used in the diagnosis and treatment of cancer. However, for a photosensitizer to be effectively used in PDT, it must meet many important criteria. Here are the key conditions that such a photosensitizer should meet:

1. Selective accumulation in cancer tissue: One of the main factors influencing the effectiveness of PDT is the ability of the photosensitizer to selectively accumulate in

cancer cells. An effective photosensitizer should preferentially accumulate in tumor tissue, minimizing absorption in healthy structures.

2. Availability and chemical properties: It is important that the photosensitizer is readily available as a pure compound and its exact chemical properties, such as reactivity and stability, should be thoroughly tested and documented. This ensures control over the synthesis process and guarantees safety and effectiveness of use.
3. No phototoxic effect in healthy tissue: It is extremely important that the photosensitizer does not cause a phototoxic effect in healthy tissue. This means that its effect should only be controlled and activated in the presence of light in the affected area.
4. High absorption coefficient in the range of 600–800 nm: the effectiveness of PDT is related to the absorption of light by the photosensitizer in the appropriate spectral range, most often in the range of 600–800 nm.
5. No overlap of absorption bands with other tissue components: The absorption band of the photosensitizer should not overlap the absorption bands of endogenous pigments such as melanin or hemoglobin, or the absorption bands of water in the near-infrared region. This avoids disruptions in the action of the photosensitizer.
6. Efficient production of $^1O_2$ or radicals: the photosensitizer should have a high quantum yield of $^1O_2$.
7. Optimal pharmacokinetic properties: the photosensitizer and its photoproducts should have appropriate pharmacokinetic properties, such as degradation, excretion, and bioavailability to ensure the effectiveness of the therapy.
8. Low side effects and easy elimination from the body: the therapeutic value of a photosensitizer is greater if it does not cause significant side effects and is easily eliminated from the body to avoid long-term toxic effects after PDT therapy.

Figure 2 shows characteristics of an ideal photosensitizer.

Meeting these criteria is crucial for the effectiveness and safety of using photosensitizers in photodynamic therapy, especially in the context of cancer treatment [158–163]. The historical development of photosensitizers in three subsequent generations is important in the context of PDT, bringing hope for further increasing its effectiveness and minimizing possible side effects.

First-generation PSs demonstrated effective PDT in the treatment of glioma [164,165] as well as limitations in the treatment of glioma resulting from their structure. First-generation PSs are limited not only by their low therapeutic effectiveness and low quantum yield of singlet oxygen [165]; first-generation PSs exhibited several limiting features that hampered their usefulness in the treatment of glioma. Therefore, it was justified to develop second-generation photosensitizers to increase the effectiveness of photodynamic therapy for glioma. Photosensitizers of the second generation have more efficient ROS production and enhanced tumor selectivity with limited adverse effects [166].

The first generation of photosensitizers includes porfimer sodium and hematoporphyrin-pioneered PDT. Although they are effective, they have limitations related to selectivity, and side effects in healthy tissues have been significant. In response to these limitations, second-generation photosensitizers were developed, such as derivatives of chlorin, bacteriochlorins, and phthalocyanines. They are characterized by stronger absorption in the deep red region of the spectrum, which allows for activation in deeper tissue and a more targeted effect on cancerous areas. This significantly increased the effectiveness of PDT. However, the greatest progress has been made in the case of third-generation photosensitizers. By combining photosensitizers with biomolecules or encapsulating them in carriers, it has become possible to significantly increase the selectivity of PDT towards cancer cells. These photosensitizers can target specific molecules or structures in cancer cells to overcome side effects [167–170]. In particular, 5-ALA has gained recognition as a beneficial photosensitizer in gliomas [171–175]. As a porphyrin precursor, 5-ALA is metabolized through the heme biosynthetic pathway to produce protoporphyrin IX (PpIX), a tumor-selective photosensitizer [98,176,177]. The high tumor selectivity for fluorescent protoporphyrin IX (PpIX) accumulation after systemic administration of 5-ALA enables intraoperative fluorescence guidance [178–181]. The advantages of 5-ALA over other photosensitizers are compelling. Numerous studies highlight the effectiveness of 5-ALA-based PDT in the treatment of gliomas, achieving cytotoxicity levels up to 80% in vitro and high tumor specificity [172,173,182–185]. PDT using 5-ALA has been shown to be safe at doses of 90 mg/kg or less followed by irradiation of rat brains with light intensity of 100 J/cm^2 [186].

Figure 2. Characteristics of an ideal photosensitizer.

5. Application

5.1. In Vivo Photodynamic Therapy

5.1.1. In Human

Perria and colleagues were the first to use PDT for glioma. Many neurosurgeons began treating glioma patients with PDT [187–193]. Fluorescence-guided tumor resection followed by PDT has been observed to prolong mean survival in patients diagnosed with glioma [194–197]. One of the first randomized, controlled trials of PDT in the treatment of malignant gliomas, conducted by Muller and Willson, included 43 glioma resections followed by PDT with photophrin, as well as a control group of 34 patients who underwent tumor resection without PDT. Median survival was 11 months in the treated group and 8 months in the control group. An increase in median survival after PDT by 38% and a higher 6-month survival rate in the treated group were demonstrated [31,196,198]. In the phase II study conducted by Kostron et al., after obtaining informed consent, 26 patients diagnosed with recurrent WHO grade IV glioma were treated with PDT using the PS Foscan (biolitec Pharma, Jena, Germany). Before treatment, all cancers were progressing and standard therapeutic options (irradiation, chemotherapy) had been exhausted. After aggressive fluorescence-guided resection (macroscopically completed in 75% of cases), intraoperative PDT was used. The median time to progression after surgery was 6 months, the median survival was 8.5 months, and the 2-year survival rate was 15%. Compared to matched controls, significantly improved survival outcomes were demonstrated in the treated group [196,199].

One study assessed the effect of photodynamic therapy and a monoclonal antibody against the vascular endothelial adhesion molecule on glioma growth in vivo. Seven days after implantation of glioma cells, adult male rats with glioma were randomly assigned to study groups as follows: (1st group) PDT + anti-vascular endothelial adhesion molecule monoclonal antibody, (2nd group) PDT, (3rd group) anti-vascular endothelial adhesion molecule monoclonal antibody, and (4th group) control. Eight days after implantation, the photosensitizer, hematoporphyrin monomethyl ether was administered, followed by PDT. In the following days, from day 8 to day 16, anti-vascular endothelial adhesion molecule monoclonal antibody was administered intravenously every other day. After 21 days, five rats were selected to be sacrificed and examined. The survival and tumor volume of the remaining 10 rats in each group were recorded. In the 3rd group after PDT, inhibition of tumor growth (67.2%) was noticed and prolongation of survival (89.3%) was observed. These effects were even more pronounced in the PDT + monoclonal antibody against vascular endothelial adhesion molecule group. It was found that PDT in combination with a monoclonal antibody against the vascular endothelial adhesion molecule effectively inhibits glioma growth and prolongs survival [200].

In 2013, Muragaki and colleagues conducted a study to evaluate the effectiveness of PDT using sodium talaporfin with the use of a 664-nm semiconductor laser. These studies were conducted in patients with primary malignant parenchymal brain tumors. Twenty-seven people were included with recurrent glioma and received a single dose of PDT with talaporfin sodium before surgery. After surgery, the area after resection were irradiated with a semiconductor laser with a wavelength of 664 nm between 22 and 27 h after PS delivery. This study was performed on 22 patients. A 12-month survival (more than 95%), 6-month progression-free survival (91%), and 6-month local progression-free survival (91%) were noticed. Incorporating intraoperative PDT into a combined treatment strategy may have a positive impact on average survival, especially in patients with newly diagnosed glioma [201].

A randomized, controlled phase III trial by the Eljamel Group evaluated fluorescence-guided resection using 5-ALA and Photofrin in repetitive PDT where 27 patients were recruited. There were 13 people in the study group with an average survival of 52.8 weeks and 14 people in the control group with an average survival of 24.6 weeks. There were no differences in complications and hospital stays between these groups. The average time to cancer progression was 8.6 months in the study group compared to 4.8 months in the control group. Therefore, it was shown that 5-ALA and Photofrin fluorescence-guided resection and repeated PDT provided a significant improvement in survival without additional risk to glioma patients [202]. A study by V. Turubanova and her team represents a significant step in understanding the potential of PDT in the context of murine glioma and fibrosarcoma. They used PSs such as a mixture of di-, tri-, and tetrasubstituted fractions of aluminum phthalocyanine and photodithazine (bis-N-methylglucamine salt of chlorin e6) in the induction of cancer cell death in response to light irradiation with a wavelength of 615 to 635 nm at a light dose of 20 J/cm^2.

In Parkinson's disease, PS accumulates mainly in the endoplasmic reticulum and the Golgi apparatus. Parkinson's disease PDT-induced cell death was inhibited only by zVAD-fmk. The most interesting aspect of this study is that dying tumor cells induced by PDT using both PS and PD emitted specific molecular signals such as calreticulin, HMGB1, and ATP. Moreover, these cells were bone marrow-derived dendritic cells (BMDCs). Importantly, dendritic cells matured, became activated, and began to secrete interleukin 6 (IL-6) [203].

A study by Sheng-Li Hu and colleagues contributed to research on the therapeutic potential of PDT in the context of the treatment of C6 glioma, which is important due to its aggressive nature in the context of brain tumors. The aim of this study was to thoroughly understand the effect of PDT on the homeostasis of calcium (Ca^{2+}) and potassium (K^+) ions in C6 glioma cells and what mechanisms influence the survival of these cells.

This study identified five experimental groups of C6 glioma cells:

1. A control group that did not undergo any form of treatment or radiotherapy.
2. A group in which cells were treated with a hematoporphyrin derivative (HpD) at a concentration of 10 mg/L but not irradiated with light.
3. PDT group in which cells were treated with HpD (10 mg/L) and irradiated with light.
4. PDT&CNQX group, in which cells were treated with HpD (10 mg/L) and the AMPA glutamate receptor antagonist, i.e., CNQX (at a concentration of 50 mol/L), and then subjected to PDT.
5. HpD&CNQX group in which cells were treated with HpD (10 mg/L) and CNQX (50 mol/L) but not irradiated with light.

The research results clearly showed that PDT causes a significant inflow of calcium ions (Ca^{2+}) into C6 glioma cells and an outflow of potassium ions (K^+), which ultimately leads to the death of these cells. An interesting finding was that treatment with an AMPA receptor antagonist (CNQX) before PDT partially blocked these changes in ion homeostasis while increasing cell survival. These observations suggest that disturbances in the homeostasis of calcium and potassium ions play a key role in the mechanisms of action of PDT on C6 glioma cells. This discovery is important for further understanding the complex mechanisms of action of PDT and its role as a potential therapeutic method in the treatment of glioma and other brain tumors [204]. The study by Deng-Pan Wu and his team focused on the therapeutic potential of PDT in the treatment of glioma. Intercellular communication (GJIC) was studied, in which the Connexin (Cx)43 protein plays a key role. Intercellular communication plays an important role in transmitting signals leading to apoptosis, i.e., programmed cell death, which may increase the effectiveness of therapies, including chemotherapy and gene therapy. The study results showed that Cx43-mediated GJIC significantly increases PDT phototoxicity, both in laboratory conditions (in vitro) and in living organisms (in vivo). This finding is significant and suggests the potential of using GJIC to improve the effectiveness of PDT. Of particular interest is that GJIC with Cx43 significantly increases PDT phototoxicity in U87 glioma cells that express Cx43. Blocking Cx43 expression had a negative impact on this effect. Additionally, the presence of GJIC with Cx43 contributed to a reduction in tumor diameter and mass after PDT using the PS Photofrin. The effectiveness of PDT with GJIC with Cx43 has been linked to the activity of stress signaling pathways such as the generation of ROS, calcium, and lipid peroxide. These observations suggest that the presence of GJIC, especially involving Cx43, may significantly increase PDT phototoxicity, which is important in the context of glioma treatment. This discovery opens prospects for further development of therapeutic strategies that aim to increase Cx43 expression or enhance GJIC function with Cx43. This, in turn, may lead to an increase in the sensitivity of cancer cells to PDT and improve its effectiveness as a therapeutic method. It is also worth noting that blocking Cx43 expression or GJIC function with Cx43 may influence the development of tumor cell resistance to PDT [205]. The aim of the study by R. Kammerer and colleagues was to investigate changes in the transcriptome of human glioma cells (U87, U373) after non-lethal dynamic phototherapy (PDT) both in vitro and in vivo. The results of the study showed that after PDT, the most increased gene expression encoded proteins related to pathways activated by cellular stress and proteins involved in cell cycle arrest. This response resembled that of cancer cells to high-dose PDT. As a result, PDT affected glioma cells by activating stress response pathways, which led to the inhibition of their cell cycle. However, there were other important observations. Cancer cells after immortalized PDT significantly upregulated a number of immune-related genes, including chemokine genes (CXCL2, CXCL3, IL8/CXCL8) as well as genes encoding interleukin-6 (IL-6) and its receptor (IL6R). These genes can stimulate inflammatory responses. Of note, IL-6 and IL6R may also influence tumor growth. Therefore, these results suggest that PDT may support the immune response against the tumor even if it is unable to completely eliminate it through phototoxic mechanisms alone. However, there is also the potential to stimulate autocrine loops that promote tumor growth, as seen in increased expression of IL-6 and its receptor. His finding suggests that despite the beneficial effects of PDT on the immune system, there may also be a risk of stimulation of tumor growth by nonlethal PDT.

Therefore, the study highlights the complexity of PDT interactions with tumor cells and the need for further research on optimal therapeutic strategies and control of inflammatory responses in the context of PDT treatment of glioma [206]. The study by Fisher and colleagues represents an important step in PDT research in the context of treating glioma, an aggressive brain tumor. The aim of the study was to determine whether mild hypothermia could influence the effectiveness of PDT by increasing the killing of glioma tumor cells while protecting normal neuronal structures. The study consisted of in vitro studies conducted on neuronal cells and in vivo studies conducted on rat models. The results of in vitro studies showed that hypothermia significantly increases the survival of neuronal cells after the use of PDT. In vivo studies in rats confirmed that hypothermia has a protective effect on neural structures after PDT, as reflected in T_2 mapping results that showed a reduction in the volume of edema and inflammation in the brain. One of the key results was an increase in protoporphyrin IX (PpIX) fluorescence in brain tumors after hypothermia. This, in turn, had a beneficial effect on the survival of rats after PDT. Histological and immunohistochemical analysis showed that hypothermia was an effective method of protecting normal brain structures during PDT. The conclusions from this study suggest that mild hypothermia may significantly improve the effectiveness of photodynamic therapy in the treatment of glioma. Hypothermia may protect neural tissues and simultaneously increase the phototoxicity of PDT, leading to the longer survival of rats after treatment. This discovery is of great importance and opens the prospects for further research on the use of hypothermia as an effective therapeutic strategy in the treatment of glioma [207]. Zhang's study used cell membranes to encapsulate indocyanine green (ICG) nanoparticles (SLNP/ICG), referred to as SLNP/ICG@M, for targeted glioma PDT. Moreover, SLNP/ICG@M produce a large amount of ROS under NIR irradiation. SLNP/ICG@M with NIR irradiation can activate the mitochondrial-mediated apoptosis pathway [208]. It is known that high concentrations of cellular glutathione (GSH) in cancer cells can reduce the ability of PDT to selectively destroy the tumor, so it is necessary to find a way to improve the therapeutic ratio of PDT in brain tumors. In a study by F Jiang et al., PDT using Photofrin as a photosensitizer combined with administration of buthionine sulfoximine (BSO), an agent that depletes glutathione levels in BSO cells, was performed in male intracerebral U87 and healthy Fisher rats. In tumor-bearing U87 rats, in vivo treatment with the PDT-BSO combination showed significantly greater tumor necrosis than individual treatment [209]. In the study by Yi et al., 24 rats with subcutaneously implanted C6 rat glioma of similar size were randomly divided into 3 groups: group 1—receiving 5-ALA-PDT, group 2—laser irradiation, and group 3—sham procedures but no treatment. Compared with groups 2 and 3, the volume of tumor grafts was significantly reduced ($p < 0.05$), MVD was significantly reduced ($p < 0.001$), and cell necrosis was significantly increased in group 1. The underlying mechanism may involve increasing cell necrosis but not inducing cellular apoptosis, which may result from the destruction of tumor micro-vessels [210]. Clinical trials have shown that PDT significantly increases the median survival of patients with gliomas. Experimental studies have shown that increasing the optical energy and PS dose leads to an increase in the volume of tumor necrosis. However, increasing the dose of light delivered to the tumor may increase the risk of causing permanent neurological deficits. In the study by Zheng et al., the neuroregenerative effect of atorvastatin on PDT was examined. However, atorvastatin significantly reduced PDT-induced cell damage. To further investigate the mechanisms underlying atorvastatin-mediated reduction in functional deficits, the effects of atorvastatin on angiogenesis and synaptogenesis were examined. Atorvastatin was found to significantly induce angiogenesis and synaptogenesis in PDT-damaged brain tissue [211]. The effectiveness of PDT is known to be induced by ROS production, dependent on hypoxia. Moreover, hypoxia activates sodium hydrogen exchanger 1 (NHE1), which is an essential component of tumor progression and metastasis. The study by Hou et al. showed that PDT significantly inhibited primary tumor growth. These results suggest that PDT with DHA may increase total ROS levels to attenuate glioma invasion and migration by suppressing NHE1 expression. The study by Kuiyuan Hou et al. showed that PDT significantly inhibited primary tumor growth, whereas PDT in synergy with DHA also inhibited recurrent tumors and improved overall survival by regulating the ROS-NHE1 axis. No visible side effects were observed. These results suggest that PDT with DHA may increase total ROS levels to attenuate GC invasion and migration by suppressing NHE1 expression [212].

In the study by Terzis et al., the effect of PDT on glioma cells (GaMg and U-251 Mg) was checked. Directional cell migration and spheroid growth were determined for both cell lines exposed to increasing laser output power (15–35 J/cm^2) at Photosan-3 photosensitizer concentrations of 5 and 7 micrograms/mL. This effect occurred within the first 4 days after exposure to the drug. Spheroids from both cell lines also showed drug dose- and laser-power-dependent growth inhibition that became apparent after a 6-day lag period. During this period, the outer layers of the spheroid cells disintegrated. The remaining spheroid tissue was unable to migrate and regrow under the highest laser energy (30–35 J/cm^2, 5 and 7 micrograms/mL Photosan-3). However, these spheroids have demonstrated the ability to invade when confronted with normal brain cells aggregated in vivo [213]. Table 1 presents examples of clinical studies with some PSs in glioma.

Table 1. Characteristics of selected clinical studies (with PS applied, excitation wavelength and treatment window).

Article	Photosensitizer	Excitation Wavelength (nm)	Treatment Window
[214]	Porfimer Sodium	630	48–150 h
[215]			
[216]			
[217]			
[218]			
[219]	Hematoporphyrin derivative [HpD]	408, 510, 630	24–48 h
[218]			
[220]	Dihematoporphyrin ether [DHE]	395, 630	24–72 h
[221]	5-Aminolevulinic Acid	410, 510, 635	4–8 h
[222]			
[218]			
[218]	Talaporfin sodium	664	12–26 h
[223]	Temoporfin [m-THPC; m-tetrahydroxyphenylchlorin]	652	48–110 h
[218]			
[224]	Boronated protoporphyrin [BOPP]	630	24 h
[218]			
[214]	Benzoporphyrin derivative [BPD]	680–690	15–30 min
[225]			
[226]			

5.1.2. In Animal Model

The study by Han-Wen Guo and colleagues used a low fluence rate and implemented long-term photodynamic therapy (PDT) in a mouse model of human glioma using an organic light-emitting diode. A single dose of 5-aminolevulinic acid (ALA) was used as a photosensitizer. Tumor volume was monitored by bioluminescence imaging and animal survival time was recorded. It was found that in animals with comparable tumor volume before and immediately after irradiation, the mean overall survival of mice subjected to PDT was 40.5 $\pm$ 9.2 days, which significantly exceeded the survival time of control mice (26.0 $\pm$ 2.0 days) [227].

Another study assessed the effects of photodynamic therapy on glioma and brain tissue. Following clinically relevant doses of PDT, reductions in tumor volume, glioma cell proliferative activity, and vascular endothelial growth factor expression were observed in the tumor area and adjacent brain tissue for 7 days. Twenty athymic mice underwent implantation of glioma cells. Fifteen mice were administered Photofrin intraperitoneally on day 6 after tumor implantation and then subjected to laser therapy with different optical doses (40 J/cm^2, 80 J/cm^2, and 120 J/cm^2) after 24 h post Photofrin injection. The remaining five tumor-bearing mice served as controls. All animals were sacrificed 14 days after tumor implantation. It was found that the tumor volume in the group of mice receiving 80 or 120 J/cm^2 was significantly smaller than in the control group. Vascular endothelial growth factor immunoreactivity in adjacent brain tissue increased significantly in mice treated with 120 J/cm^2 PDT compared to the control group, as well as in mice treated with Photofrin and lower optical doses. No significant differences in glioma cell proliferation and vascular endothelial growth factor expression in the tumor area were observed between groups. These results suggest that PDT is effective in tumor shrinkage, especially with higher light doses, and PDT-induced vascular endothelial growth factor expression in adjacent brain tissue may be associated with tumor recurrence. Therefore, combining PDT with antiangiogenic drugs may be an effective strategy for the treatment of glioma [228]. In order to evaluate the effectiveness of the synthesized nanoparticles dedicated to PDT glioma, athymic mice were bred. Glioma was induced by subcutaneous injection of U87-MG cells. PDT was performed 15 days after implantation. CPN was administered intravenously and intratumorally. Mice treated with a single i.t. or intravenous CPN administration showed strong tumor growth inhibition for 10 days after PDT treatment. Tumors from the PBS control or nonirradiated CPN groups reached more than twice their initial volume [229].

Human monocyte cells (THP-1) and mouse monocytes isolated from bone marrow (mBMDM) were used as covert CPN carriers to penetrate glioma spheroids and an orthotopic tumor model. The success of PDT using this cell-mediated targeting strategy was determined by its effect on spheroids. CPNs did not affect monocyte viability in the absence of light and did not show nonspecific release after cell loading [230].

6. PDT Glioma Current Clinical Research

Photodynamic therapy is approved by the US Food and Drug Administration (FDA) for the treatment of premalignant diseases and malignancies such as actinic keratosis, Barrett's esophagus, esophageal cancer, and non-small cell lung cancer [171]. Many of the novel concepts and strategies for using PDT in the treatment of glioma are in the in vitro experimental phase, still requiring extensive evidence of effectiveness studies before clinical application [9]. A study by Vermandel et al. examined the safety and efficacy after intraoperative treatment of glioma with photodynamic therapy (PDT) after administration of 5-ALA acid and maximal resection in 10 patients with newly diagnosed glioma. There were no serious adverse events, and it was shown that this strategy could help reduce the risk of recurrence by targeting residual tumor cells in the resection cavity, as the 12-month recurrence-free survival rate was 60% [231]. An ongoing study in Germany is evaluating stereotactic biopsy followed by 5-ALA-based stereotactic PDT and the feasibility of 5-ALA in stereotactic interstitial PDT in a subset of adult glioma patients [9]. The only fluorescence-guided glioma surgery agent approved by the FDA [232] is 5-ALA.

In Scotland, over 365 brain treatments were performed in patients using 5-ALA and Photofrin intracavitary PDT with a balloon diffuser and a diode laser with a wavelength of 630 nm. Of these treatment cases, 143 had no adverse effects (Figure 3). Adverse events were reported in seven patients: three patients developed deep vein thrombosis (DVT), two patients experienced skin photosensitivity due to lack of light protection, two patients experienced cerebral edema after PDT, one patient experienced skin necrosis and cerebrospinal fluid leakage, and in one patient the balloon diffuser ruptured due to poor catheter attachment [124,170].

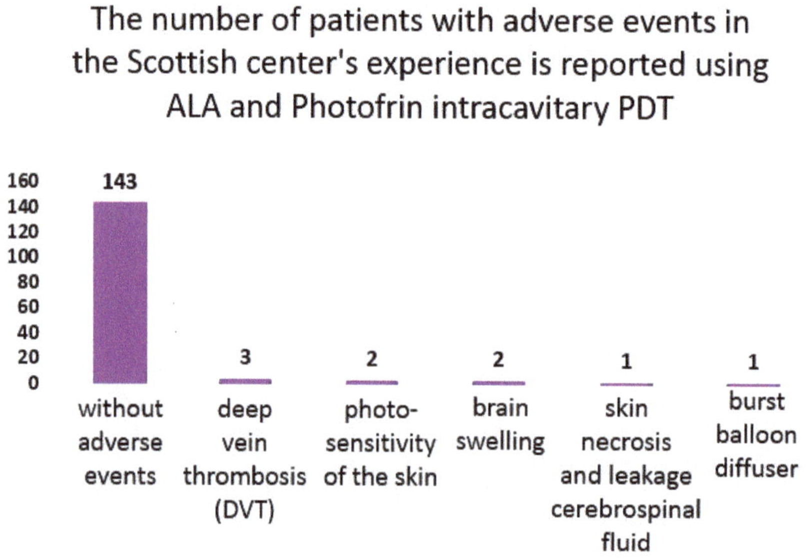

Figure 3. Adverse events in the Scottish center study using ALA and Photofrin intracavitary PDT with a balloon diffuser and a 630 nm diode laser on brain tissue.

These side effects are no different from those occurring during other therapies [233,234]. Research conducted on 20 patients treated with HPD and 630 nm light revealed five expected skin photosensitivities, none of which were serious adverse events [233–235]. Only 2 of 136 patients treated with HPD-PDT in Australia (2%) reported excessive sunburn associated with skin photosensitivity [236]. In both cases, they did not follow the recommended instructions. The main side effect associated with treatment is short-term hypersensitivity of the skin and retina to light after the administration of a PS [196].

7. Conclusions

Research on PDT in the treatment of gliomas is a fascinating area of medical research that opens new perspectives in the fight against this aggressive form of brain cancer. Previous and current in vivo studies and patient clinical experiences show promising results with PDT as an effective and increasingly popular therapeutic option in the treatment of gliomas. The most important finding from this study is the significant improvement in median survival for patients who received PDT as part of their treatment. Prolonged progression-free survival has been observed in many cases, which is an important step in the fight against this serious disease. These results suggest that PDT may provide a valuable alternative to traditional glioma treatments such as surgery, radiotherapy, and chemotherapy, which are often associated with significant side effects and limited effectiveness. Additionally, research on PDT has shown that this therapy can stimulate the body's immune response against cancer cells. This finding is particularly promising because it highlights the potential of PDT as a tool to induce a defense response against gliomas, which may further aid in tumor elimination. It is worth emphasizing that the results of in vivo studies on animal models and clinical studies on patients unanimously suggest the benefits of using PDT in the treatment of gliomas. This is important evidence of the promising nature of this therapy and its potential to change the standards of treatment for this difficult disease. Although the results of these studies are encouraging, further research is necessary to optimize PDT parameters, including the selection of appropriate photosensitizers, light parameters, and combination treatment strategies. Moreover, a key challenge is to adapt PDT to the individual patient's case, taking into account differences in disease stage and molecular caesuras. These conclusions emphasize the importance of continuing research on PDT as a promising therapeutic strategy for patients suffering from gliomas. Further development of this field of medicine may bring significant benefits to patients in the form of improved quality and length of survival, as well as influence future standards of treatment of this difficult and fatal disease.

Author Contributions: Conceptualization, D.A., A.P., A.M., K.D., M.K.-O., A.B., A.K.-K. and D.B.-A.; methodology, D.A., A.P., A.M., K.D., M.K.-O., A.B., A.K.-K. and D.B.-A.; software, D.A., A.P., A.M., K.D., M.K.-O., A.B., A.K.-K. and D.B.-A.; validation, D.A., A.P., A.M., K.D., M.K.-O., A.B., A.K.-K. and D.B.-A.; formal analysis, D.A., A.P., A.M., K.D., M.K.-O., A.B., A.K.-K. and D.B.-A.; resources, D.A., A.P., A.M., K.D., M.K.-O., A.B., A.K.-K. and D.B.-A.; data curation, D.A., A.P., A.M., K.D., M.K.-O., A.B., A.K.-K. and D.B.-A.; writing—original draft preparation, D.A., A.P., A.M., K.D., M.K.-O., A.B., A.K.-K. and D.B.-A.; writing—review and editing, D.A., A.P., A.M., K.D., M.K.-O., A.B., A.K.-K. and D.B.-A.; visualization, D.A., A.P., A.M., K.D., M.K.-O., A.B., A.K.-K. and D.B.-A.; supervision, D.A.; funding acquisition, D.A., A.B. and A.K.-K. All authors have read and agreed to the published version of the manuscript.

Funding: This research received no external funding.

Conflicts of Interest: The authors declare no conflicts of interest.

References

1. Xu, S.; Tang, L.; Li, X.; Fan, F.; Liu, Z. Immunotherapy for glioma: Current management and future application. *Cancer Lett.* **2020**, *476*, 1–12. [CrossRef] [PubMed]
2. Sioutas, G.; Nikova, A.; Birbilis, T. Risk factors for pediatric glioma. *Folia Med.* **2022**, *64*, 566–571. [CrossRef] [PubMed]
3. Poff, A.; Koutnik, A.P.; Egan, K.M.; Sahebjam, S.; D'Agostino, D.; Kumar, N.B. Targeting the Warburg effect for cancer treatment: Ketogenic diets for management of glioma. *Semin. Cancer Biol.* **2019**, *56*, 135–148. [CrossRef] [PubMed]
4. Hakar, M.H.; Wood, M.D. Updates in Pediatric Glioma Pathology. *Surg. Pathol. Clin.* **2020**, *13*, 801–816. [CrossRef] [PubMed]
5. Gisina, A.; Kholodenko, I.; Kim, Y.; Abakumov, M.; Lupatov, A.; Yarygin, K. Glioma Stem Cells: Novel Data Obtained by Single-Cell Sequencing. *Int. J. Mol. Sci.* **2022**, *23*, 14224. [CrossRef] [PubMed]
6. Miller, A.M.; Shah, R.H.; Pentsova, E.I.; Pourmaleki, M.; Briggs, S.; Distefano, N.; Zheng, Y.; Skakodub, A.; Mehta, S.A.; Campos, C.; et al. Tracking tumour evolution in glioma through liquid biopsies of cerebrospinal fluid. *Nature* **2019**, *565*, 654–658. [CrossRef] [PubMed]
7. Li, J.; Meng, Q.; Zhou, X.; Zhao, H.; Wang, K.; Niu, H.; Wang, Y. Gospel of malignant Glioma: Oncolytic virus therapy. *Gene* **2022**, *818*, 146217. [CrossRef]
8. Śledzińska, P.; Bebyn, M.G.; Furtak, J.; Kowalewski, J.; Lewandowska, M.A. Prognostic and Predictive Biomarkers in Gliomas. *Int. J. Mol. Sci.* **2021**, *22*, 10373. [CrossRef]
9. Bhanja, D.; Wilding, H.; Baroz, A.; Trifoi, M.; Shenoy, G.; Slagle-Webb, B.; Hayes, D.; Soudagar, Y.; Connor, J.; Mansouri, A. Photodynamic Therapy for Glioblastoma: Illuminating the Path toward Clinical Applicability. *Cancers* **2023**, *15*, 3427. [CrossRef]
10. Ghantasala, S.; Gollapalli, K.; Epari, S.; Moiyadi, A.; Srivastava, S. Glioma tumor proteomics: Clinically useful protein biomarkers and future perspectives. *Expert Rev. Proteom.* **2020**, *17*, 221–232. [CrossRef]
11. Tan, A.C.; Ashley, D.M.; López, G.Y.; Malinzak, M.; Friedman, H.S.; Khasraw, M. Management of glioblastoma: State of the art and future directions. *CA Cancer J. Clin.* **2020**, *70*, 299–312. [CrossRef]
12. Stupp, R.; Mason, W.P.; van den Bent, M.J.; Weller, M.; Fisher, B.; Taphoorn, M.J.B.; Belanger, K.; Brandes, A.A.; Marosi, C.; Bogdahn, U.; et al. Radiotherapy plus Concomitant and Adjuvant Temozolomide for Glioblastoma. *N. Engl. J. Med.* **2005**, *352*, 987–996. [CrossRef] [PubMed]
13. Tsitlakidis, A.; Aifantis, E.C.; Kritis, A.; Tsingotjidou, A.S.; Cheva, A.; Selviaridis, P.; Foroglou, N. Mechanical properties of human glioma. *Neurol. Res.* **2020**, *42*, 1018–1026. [CrossRef] [PubMed]
14. Guo, X.; Sui, R.; Piao, H. Exosomes-mediated crosstalk between glioma and immune cells in the tumor microenvironment. *CNS Neurosci. Ther.* **2023**, *29*, 2074–2085. [CrossRef] [PubMed]

15. Pienkowski, T.; Kowalczyk, T.; Garcia-Romero, N.; Ayuso-Sacido, A.; Ciborowski, M. Proteomics and metabolomics approach in adult and pediatric glioma diagnostics. *Biochim. Biophys. Acta Rev. Cancer* **2022**, *1877*, 188721. [CrossRef]
16. Wang, L.M.; Englander, Z.K.; Miller, M.L.; Bruce, J.N. Malignant Glioma. In *Human Brain and Spinal Cord Tumors: From Bench to Bedside. Volume 2: The Path to Bedside Management*; Springer: Cham, Switzerland, 2023; Volume 1405, pp. 1–30.
17. Campos, B.; Olsen, L.R.; Urup, T.; Poulsen, H.S. A comprehensive profile of recurrent glioblastoma. *Oncogene* **2016**, *35*, 5819–5825. [CrossRef] [PubMed]
18. Da Ros, M.; De Gregorio, V.; Iorio, A.L.; Giunti, L.; Guidi, M.; De Martino, M.; Genitori, L.; Sardi, I. Glioblastoma Chemoresistance: The Double Play by Microenvironment and Blood-Brain Barrier. *Int. J. Mol. Sci.* **2018**, *19*, 2879. [CrossRef]
19. Ostrom, Q.T.; Bauchet, L.; Davis, F.G.; Deltour, I.; Fisher, J.L.; Langer, C.E.; Pekmezci, M.; Schwartzbaum, J.A.; Turner, M.C.; Walsh, K.M.; et al. The epidemiology of glioma in adults: A "state of the science" review. *Neuro-Oncology* **2014**, *16*, 896–913. [CrossRef]
20. Tirapelli, L.F.; Morgueti, M.; da Cunha Tirapelli, D.P.; Bagnato, V.S.; Ferreira, J.; Neto, F.S.; Peria, F.M.; Oliveira, H.F.; Junior, C.G. Apoptosis in glioma cells treated with PDT. *Photomed. Laser Surg.* **2011**, *29*, 305–309. [CrossRef]
21. Wang, T.J.C.; Mehta, M.P. Low-Grade Glioma Radiotherapy Treatment and Trials. *Neurosurg. Clin. N. Am.* **2019**, *30*, 111–118. [CrossRef]
22. Terashima, K.; Ogiwara, H. Pediatric Glioma. *No Shinkei Geka* **2021**, *49*, 640–646. [PubMed]
23. Davis, M.E. Epidemiology and Overview of Gliomas. *Semin. Oncol. Nurs.* **2018**, *34*, 420–429. [CrossRef] [PubMed]
24. Kessel, D.; Vicente, M.G.; Reiners, J.J., Jr. Initiation of apoptosis and autophagy by photodynamic therapy. *Autophagy* **2006**, *2*, 289–290. [CrossRef] [PubMed]
25. Horne, T.K.; Cronje, M.J. Cancer Tissue Classification, Associated Therapeutic Implications and PDT as an Alternative. *Anticancer Res.* **2017**, *37*, 2785–2807.
26. Zhang, Q.; Li, L. Photodynamic combinational therapy in cancer treatment. *J. BUON* **2018**, *23*, 561–567.
27. Eljamel, M.S. Brain PDD and PDT unlocking the mystery of malignant gliomas. *Photodiagn. Photodyn. Ther.* **2005**, *1*, 303–310. [CrossRef]
28. Pichlmeier, U.; Bink, A.; Schackert, G.; Stummer, W. Resection and survival in glioblastoma multiforme: An RTOG recursive partitioning analysis of ALA study patients. *Neuro-Oncology* **2008**, *10*, 1025–1034. [CrossRef]
29. Kaneko, S. A current overview: Photodynamic diagnosis and photodynamic therapy using ALA in neurosurgery. *J. Jpn. Soc. Laser Surg. Med.* **2008**, *29*, 135–146. [CrossRef]
30. Kostron, H.; Grunert, V. Photodynamic therapy of malignant brain tumors. *Wien. Klin. Wochenschr.* **1987**, *99*, 389–392. [PubMed]
31. Muller, P.J.; Wilson, B.C. Photodynamic therapy of malignant brain tumours. *Can. J. Neurol. Sci.* **1990**, *17*, 193–198. [CrossRef] [PubMed]
32. Moan, J.; Peng, Q. An outline of the hundred-year history of PDT. *Anticancer Res.* **2003**, *23*, 3591–3600.
33. Kabingu, E.; Oseroff, A.R.; Wilding, G.E.; Gollnick, S.O. Enhanced systemic immune reactivity to a Basal cell carcinoma associated antigen following photodynamic therapy. *Clin. Cancer Res.* **2009**, *15*, 4460–4466. [CrossRef]
34. Gollnick, S.O.; Brackett, C.M. Enhancement of anti-tumor immunity by photodynamic therapy. *Immunol. Res.* **2010**, *46*, 216–226. [CrossRef]
35. Castano, A.P.; Mroz, P.; Hamblin, M.R. Photodynamic therapy and antitumour immunity. *Nat. Rev. Cancer* **2006**, *6*, 535–545. [CrossRef] [PubMed]
36. Garg, A.D.; Nowis, D.; Golab, J.; Agostinis, P. Photodynamic therapy: Illuminating the road from cell death towards anti-tumour immunity. *Apoptosis* **2010**, *15*, 1050–1071. [CrossRef]
37. Ritz, R.; Wein, H.T.; Dietz, K.; Schenk, M.; Roser, F.; Tatagiba, M.; Strauss, W.S. Photodynamic therapy of malignant glioma with hypericin: Comprehensive in vitro study in human glioblastoma cell lines. *Int. J. Oncol.* **2007**, *30*, 659–667. [CrossRef] [PubMed]
38. Hirschberg, H.; Sun, C.-H.; Tromberg, B.J.; Madsen, S. ALA- and ALA-esters-mediated phototdynamic therapy of human glioma spheroids. *J. Neuro-Oncol.* **2002**, *57*, 1–7. [CrossRef]
39. Madsen, S.J.; Sun, C.H.; Tromberg, B.J.; Wallace, V.P.; Hirschberg, H. Photodynamic therapy of human glioma spheroids using 5-aminolevulinic acid. *Photochem. Photobiol.* **2000**, *72*, 128–134. [CrossRef]
40. Angell-Petersen, E.; Spetalen, S.; Madsen, S.; Sun, C.; Peng, Q.; Carper, S.; Sioud, M.; Hirschberg, H. Influence of light fluence rate on the effects of photodynamic therapy in an orthopic rat glioma model. *J. Neurosurg.* **2006**, *104*, 109–117. [CrossRef]
41. Hirschberg, H.; Sorensen, D.; Angell-Petersen, E.; Peng, Q.; Tromberg, B.; Sun, C.; Spetalen, S.; Madsen, S. Repetive photodynamic therapy of malignant brain tumors. *J. Environ. Pathol. Toxicol. Oncol.* **2006**, *25*, 261–280. [CrossRef] [PubMed]
42. Ito, S.; Rachinger, W.; Stepp, H.; Reulen, H.J.; Stummer, W. Oedema formation in experimental photo-irradiation of brain tumors using 5-ALA. *Acta Neurochir.* **2005**, *147*, 57–65. [CrossRef] [PubMed]
43. Olzowy, B.; Hundt, C.; Stocker, S.; Bise, K.; Reulen, H.J.; Stummer, W. Photoirradiation therapy of experimental malignant glioma with 5-aminolevulinic acid. *J. Neurosurg.* **2002**, *97*, 970–976. [CrossRef] [PubMed]
44. Yasinjan, F.; Xing, Y.; Geng, H.; Guo, R.; Yang, L.; Liu, Z.; Wang, H. Immunotherapy: A promising approach for glioma treatment. *Front. Immunol.* **2023**, *14*, 1255611. [CrossRef] [PubMed]
45. Zhu, X.; Zhou, H.; Liu, Y.; Wen, Y.; Wei, C.; Yu, Q.; Liu, J. Transferrin/aptamer conjugated mesoporous ruthenium nanosystem for redox-controlled and targeted chemo-photodynamic therapy of glioma. *Acta Biomater.* **2018**, *82*, 143–157. [CrossRef] [PubMed]
46. Castano, A.P.; Demidova, T.N.; Hamblin, M.R. Mechanisms in photodynamic therapy: Part two-cellular signaling, cell metabolism and modes of cell death. *Photodiagn. Photodyn. Ther.* **2005**, *2*, 1–23. [CrossRef] [PubMed]

47. Nayak, L.; Reardon, D.A. High-grade Gliomas. *Continuum (Minneap. Minn.)* **2017**, *23*, 1548–1563. [CrossRef]
48. Shergalis, A.; Bankhead, A., 3rd; Luesakul, U.; Muangsin, N.; Neamati, N. Current Challenges and Opportunities in Treating Glioblastoma. *Pharmacol. Rev.* **2018**, *70*, 412–445. [CrossRef]
49. Cruz, J.V.R.; Batista, C.; Afonso, B.H.; Alexandre-Moreira, M.S.; Dubois, L.G.; Pontes, B.; Moura Neto, V.; Mendes, F.A. Obstacles to Glioblastoma Treatment Two Decades after Temozolomide. *Cancers* **2022**, *14*, 3203. [CrossRef]
50. Lara-Velazquez, M.; Al-Kharboosh, R.; Jeanneret, S.; Vazquez-Ramos, C.; Mahato, D.; Tavanaiepour, D.; Rahmathulla, G.; Quinones-Hinojosa, A. Advances in Brain Tumor Surgery for Glioblastoma in Adults. *Brain Sci.* **2017**, *7*, 166. [CrossRef]
51. Matsumae, M.; Nishiyama, J.; Kuroda, K. Intraoperative MR Imaging during Glioma Resection. *Magn. Reson. Med. Sci.* **2022**, *21*, 148–167. [CrossRef]
52. Bush, N.A.; Chang, S.M.; Berger, M.S. Current and future strategies for treatment of glioma. *Neurosurg. Rev.* **2017**, *40*, 1–14. [CrossRef] [PubMed]
53. Sanai, N.; Berger, M. Glioma extent of resection and its impact on patient outcome. *Neurosurgery* **2008**, *62*, 753–766. [CrossRef]
54. Vives, K.P.; Piepmeier, J.M. Complications and Expected Outcome of Glioma Surgery. *J. Neuro-Oncol.* **1999**, *42*, 289–302. [CrossRef] [PubMed]
55. De Witt Hamer, P.C.; De Witt Hamer, P.C.; Klein, M.; Hervey-Jumper, S.L.; Wefel, J.S.; Berger, M.S. Functional Outcomes and Health-Related Quality of Life Following Glioma Surgery. *Neurosurgery* **2021**, *88*, 720–732. [CrossRef]
56. Aiyappa-Maudsley, R.; Chalmers, A.J.; Parsons, J.L. Factors affecting the radiation response in glioblastoma. *Neuro-Oncol. Adv.* **2022**, *4*, vdac156. [CrossRef] [PubMed]
57. Liu, P.; Liu, G.; Wang, G.; Zhou, W.; Sun, Y.; Chen, W.; Zeng, Q.; Hong, J.; Xie, Q.; Ou, L.; et al. Comparison of Dosimetric Gains Provided by Intensity-Modulated Radiotherapy, Volume-Modulated Arc Therapy, and Helical Tomotherapy for High-Grade Glioma. *BioMed Res. Int.* **2020**, *2020*, 4258989. [CrossRef]
58. Zhao, M.; Li, Y.; Lu, C.; Ding, F.; Xu, M.; Ge, X.; Li, M.; Wang, Z.; Yin, J.; Zhang, J.; et al. PGC1α Degradation Suppresses Mitochondrial Biogenesis to Confer Radiation Resistance in Glioma. *Cancer Res.* **2023**, *83*, 1094–1110. [CrossRef]
59. Barani, I.J.; Larson, D.A. Radiation Therapy of Glioblastoma. In *Current Understanding and Treatment of Gliomas*; Raizer, J., Parsa, A., Eds.; Cancer Treatment and Research; Springer: Cham, Switzerland, 2015; Volume 163.
60. Rončević, A.; Koruga, N.; Soldo Koruga, A.; Rončević, R.; Rotim, T.; Šimundić, T.; Kretić, D.; Perić, M.; Turk, T.; Štimac, D. Personalized Treatment of Glioblastoma: Current State and Future Perspective. *Biomedicines* **2023**, *11*, 1579. [CrossRef]
61. Raisa, N.; Marhaendraputro, E.A. the side effects of chemotherapy in glioma. *Malang Neurol. J.* **2019**, *5*, 92–97. [CrossRef]
62. Zajączkowska, R.; Kocot-Kępska, M.; Leppert, W.; Wrzosek, A.; Mika, J.; Wordliczek, J. Mechanisms of Chemotherapy-Induced Peripheral Neuropathy. *Int. J. Mol. Sci.* **2019**, *20*, 1451. [CrossRef]
63. Nayak, R.; Mallick, B. LncRNA-associated competing endogenous RNA network analysis uncovered key lncRNAs involved in temozolomide resistance and tumor recurrence of glioblastoma. *J. Mol. Recognit.* **2023**, *36*, e3060. [CrossRef]
64. Syro, L.V.; Rotondo, F.; Camargo, M.; Ortiz, L.D.; Serna, C.A.; Kovacs, K. Temozolomide and Pituitary Tumors: Current Understanding, Unresolved Issues, and Future Directions. *Front. Endocrinol.* **2018**, *9*, 318. [CrossRef] [PubMed]
65. Aasland, D.; Götzinger, L.; Hauck, L.; Berte, N.; Meyer, J.; Effenberger, M.; Schneider, S.; Reuber, E.E.; Roos, W.P.; Tomicic, M.T.; et al. Temozolomide Induces Senescence and Repression of DNA Repair Pathways in Glioblastoma Cells via Activation of ATR-CHK1, p21, and NF-κB. *Cancer Res.* **2019**, *79*, 99–113. [CrossRef] [PubMed]
66. Wu, W.; Klockow, J.L.; Zhang, M.; Lafortune, F.; Chang, E.; Jin, L.; Wu, Y.; Daldrup-Link, H.E. Glioblastoma multiforme (GBM): An overview of current therapies and mechanisms of resistance. *Pharmacol. Res.* **2021**, *171*, 105780. [CrossRef] [PubMed]
67. Chamberlain, M.C. Temozolomide: Therapeutic limitations in the treatment of adult high-grade gliomas. *Expert Rev. Neurother.* **2010**, *10*, 1537–1544. [CrossRef] [PubMed]
68. Tomar, M.S.; Kumar, A.; Srivastava, C.; Shrivastava, A. Elucidating the mechanisms of Temozolomide resistance in gliomas and the strategies to overcome the resistance. *Biochim. Biophys. Acta Rev. Cancer* **2021**, *1876*, 188616. [CrossRef]
69. Afonso, M.; Brito, M.A. Therapeutic Options in Neuro-Oncology. *Int. J. Mol. Sci.* **2022**, *23*, 5351. [CrossRef]
70. Taal, W.; Bromberg, J.E.; van den Bent, M.J. Chemotherapy in glioma. *CNS Oncol.* **2015**, *4*, 179–192. [CrossRef]
71. Nikolova, T.; Roos, W.P.; Krämer, O.H.; Strik, H.M.; Kaina, B. Chloroethylating nitrosoureas in cancer therapy: DNA damage, repair and cell death signaling. *Biochim. Biophys. Acta Rev. Cancer.* **2017**, *1868*, 29–39. [CrossRef]
72. Kaina, B.; Christmann, M. DNA repair in personalized brain cancer therapy with temozolomide and nitrosoureas. *DNA Repair* **2019**, *78*, 128–141. [CrossRef]
73. Lombardi, G.; De Salvo, G.L.; Brandes, A.A.; Eoli, M.; Rudà, R.; Faedi, M.; Lolli, I.; Pace, A.; Daniele, B.; Pasqualetti, F.; et al. Regorafenib compared with lomustine in patients with relapsed glioblastoma (REGOMA): A multicentre, open-label, randomised, controlled, phase 2 trial. *Lancet Oncol.* **2019**, *20*, 110–119. [CrossRef]
74. Weller, M.; Le Rhun, E. How did lomustine become standard of care in recurrent glioblastoma? *Cancer Treat. Rev.* **2020**, *87*, 102029. [CrossRef]
75. van den Bent, M.; Eoli, M.; Sepulveda, J.M.; Smits, M.; Walenkamp, A.; Frenel, J.-S.; Franceschi, E.; Clement, P.M.; Chinot, O.; De Vos, F.; et al. INTELLANCE 2/EORTC 1410 randomized phase II study of Depatux-M alone and with temozolomide vs temozolomide or lomustine in recurrent EGFRamplified glioblastoma. *Neuro-Oncology* **2019**, *22*, 684–693. [CrossRef]
76. Xiao, Z.Z.; Wang, Z.F.; Lan, T.; Huang, W.H.; Zhao, Y.H.; Ma, C.; Li, Z.Q. Carmustine as a Supplementary Therapeutic Option for Glioblastoma: A Systematic Review and Meta-Analysis. *Front. Neurol.* **2020**, *11*, 1036. [CrossRef] [PubMed]

77. Roux, A.; Peeters, S.; Zanello, M.; Bou Nassif, R.; Abi Lahoud, G.; Dezamis, E.; Parraga, E.; Lechapt-Zalcmann, E.; Dhermain, F.; Dumont, S.; et al. Extent of resection and Carmustine wafer implantation safely improve survival in patients with a newly diagnosed glioblastoma: A single center experience of the current practice. *J. Neurooncol.* **2017**, *135*, 83–92. [CrossRef]
78. Brandes, A.A.; Bartolotti, M.; Tosoni, A.; Franceschi, E. Nitrosoureas in the Management of Malignant Gliomas. *Curr. Neurol. Neurosci. Rep.* **2016**, *16*, 13. [CrossRef] [PubMed]
79. Lillehei, K.O.; Kalkanis, S.N.; Liau, L.M.; Mydland, D.E.; Olson, J.; Paleologos, N.A.; Ryken, T.; Johnson, T.; Scullin, E.; Weller, M.; et al. Rationale and design of the 500-patient, 3-year, and prospective Vigilant ObservatIon of GlIadeL WAfer ImplaNT registry. *CNS Oncol.* **2018**, *7*, CNS08. [CrossRef] [PubMed]
80. Bethesda, M. PDQ®Adult Treatment Editorial Board PDQ Adult Central Nervous System Tumors Treatment. Available online: https://www.ncbi.nlm.nih.gov/books/NBK65982/ (accessed on 10 June 2023).
81. Addeo, R.; Caraglia, M.; De Santi, M.S.; Montella, L.; Abbruzzese, A.; Parlato, C.; Vincenzi, B.; Carraturo, M.; Faiola, V.; Genovese, M.; et al. A new schedule of fotemustine in temozolomide-pretreated patients with relapsing glioblastoma. *J. Neurooncol.* **2011**, *102*, 417–424. [CrossRef]
82. Stupp, R.; Taillibert, S.; Kanner, A.; Read, W.; Steinberg, D.; Lhermitte, B.; Toms, S.; Idbaih, A.; Ahluwalia, M.S.; Fink, K.; et al. Effect of Tumor-Treating Fields Plus Maintenance Temozolomide vs Maintenance Temozolomide Alone on Survival in Patients With Glioblastoma: A Randomized Clinical Trial. *JAMA* **2017**, *318*, 2306–2316. [CrossRef]
83. Sathornsumetee, S.; Reardon, D.A.; Desjardins, A.; Quinn, J.A.; Vredenburgh, J.J.; Rich, J.N. Molecularly targeted therapy for malignant glioma. *Cancer* **2007**, *110*, 13–24. [CrossRef]
84. Yang, K.; Wu, Z.; Zhang, H. Glioma targeted therapy: Insight into future of molecular approaches. *Mol. Cancer* **2022**, *21*, 39. [CrossRef]
85. Liu, Z.; Han, H.; He, X.; Li, S.; Wu, C.; Yu, C.; Wang, S. Expression of the Galectin-9-Tim-3 Pathway in Glioma Tissues is Associated With the Clinical Manifestations of Glioma. *Oncol. Lett.* **2016**, *11*, 1829–1834. [CrossRef] [PubMed]
86. Harris-Bookman, S.; Mathios, D.; Martin, A.M.; Xia, Y.; Kim, E.; Xu, H.; Belcaid, Z.; Polanczyk, M.; Barberi, T.; Theodros, D.; et al. Expression of LAG-3 and Efficacy of Combination Treatment With Anti-LAG-3 and Anti-PD-1 Monoclonal Antibodies in Glioblastoma. *Int. J. Cancer* **2018**, *143*, 3201–3208. [CrossRef] [PubMed]
87. Lamrani, M.; Sassi, N.; Paul, C.; Yousfi, N.; Boucher, J.L.; Gauthier, N.; Labbé, J.; Seignez, C.; Racoeur, C.; Athias, A.; et al. Tlr4/Ifnγ Pathways Induce Tumor Regression via NOS II-Dependent NO and ROS Production in Murine Breast Cancer Models. *OncoImmunology* **2016**, *5*, e1123369. [CrossRef] [PubMed]
88. Liu, F.; Huang, J.; Liu, X.; Cheng, Q.; Luo, C.; Liu, Z. CTLA-4 Correlates With Immune and Clinical Characteristics of Glioma. *Cancer Cell Int.* **2020**, *20*, 7. [CrossRef] [PubMed]
89. Schalper, K.A.; Rodriguez-Ruiz, M.E.; Diez-Valle, R.; López-Janeiro, A.; Porciuncula, A.; Idoate, M.A.; Inogés, S.; De Andrea, C.; López-Diaz De Cerio, A.; Tejada, S.; et al. Neoadjuvant Nivolumab Modifies the Tumor Immune Microenvironment in Resectable Glioblastoma. *Nat. Med.* **2019**, *25*, 470–476. [CrossRef] [PubMed]
90. Filley, A.C.; Henriquez, M.; Dey, M. Recurrent Glioma Clinical Trial, CheckMate-143: The Game is Not Over Yet. *Oncotarget* **2017**, *8*, 91779–91794. [CrossRef] [PubMed]
91. Ghouzlani, A.; Kandoussi, S.; Tall, M.; Reddy, K.P.; Rafii, S.; Badou, A. Immune Checkpoint Inhibitors in Human Glioma Microenvironment. *Front. Immunol.* **2021**, *12*, 679425. [CrossRef] [PubMed]
92. Nejo, T.; Yamamichi, A.; Almeida, N.D.; Goretsky, Y.E.; Okada, H. Tumor antigens in glioma. *Semin. Immunol.* **2020**, *47*, 101385. [CrossRef]
93. Kewal, K. Critical Overview of Targeted Therapies for Glioblastoma. *Front. Oncol.* **2018**, *8*, 419.
94. Ameratunga, M.; Pavlakis, N.; Wheeler, H.; Grant, R.; Simes, J.; Khasraw, M. Anti-angiogenic therapy for high-grade glioma. *Cochrane Database Syst. Rev.* **2018**, *2018*, CD008218. [CrossRef]
95. Ozel, O.; Kurt, M.; Ozdemir, O.; Bayram, J.; Akdeniz, H.; Koca, D. Complete response to bevacizumab plus irinotecan in patients with rapidly progressive GBM: Cases report and literature review. *J. Oncol. Sci.* **2016**, *2*, 87–94. [CrossRef]
96. Brandes, A.A.; Finocchiaro, G.; Zagonel, V.; Reni, M.; Caserta, C.; Fabi, A.; Clavarezza, M.; Maiello, E.; Eoli, M.; Lombardi, G.; et al. AVAREG: A phase II, randomized, noncomparative study of fotemustine or bevacizumab for patients with recurrent glioblastoma. *Neuro-Oncology* **2016**, *18*, 1304–1312. [CrossRef]
97. Diaz, R.J.; Ali, S.; Qadir, M.G.; De La Fuente, M.I.; Ivan, M.E.; Komotar, R.J. The role of bevacizumab in the treatment of glioblastoma. *J. Neuro-Oncol.* **2017**, *133*, 455–467. [CrossRef] [PubMed]
98. Stewart, L.A. Chemotherapy in adult high-grade glioma: A systematic review and meta-analysis of individual patient data from 12 randomised trials. *Lancet* **2002**, *359*, 1011–1018.
99. Fujii, J.; Soma, Y.; Matsuda, Y. Biological Action of Singlet Molecular Oxygen from the Standpoint of Cell Signaling, Injury and Death. *Molecules* **2023**, *28*, 4085. [CrossRef]
100. Choromańska, A.; Kulbacka, J.; Saczko, J. Terapia fotodynamiczna-założenia, mechanizm, aplikacje kliniczne. *Nowa Med.* **2013**, *20*, 26–30.
101. Fontana, L.C.; Pinto, J.G.; Vitorio, G.D.S.; Ferreira, I.; Pacheco-Soares, C.; Mamone, L.A.; Strixino, J.F. Photodynamic effect of protoporphyrin IX in gliosarcoma 9l/lacZ cell line. *Photodiagn. Photodyn. Ther.* **2022**, *37*, 102669. [CrossRef]
102. de Paula, L.B.; Primo, F.L.; Tedesco, A.C. Nanomedicine associated with photodynamic therapy for glioblastoma treatment. *Biophys. Rev.* **2017**, *9*, 761–773. [CrossRef] [PubMed]

103. Howley, R.; Chandratre, S.; Chen, B. 5-Aminolevulinic Acid as a Theranostic Agent for Tumor Fluorescence Imaging and Photodynamic Therapy. *Bioengineering* **2023**, *10*, 496. [CrossRef]
104. Chen, J.; Zhou, D.; Kang, J.; Liu, C.; Huang, R.; Jiang, Z.; Liao, Y.; Liu, A.; Gao, L.; Song, X.; et al. ER stress modulates apoptosis in A431 cell subjected to EtNBSe-PDT via the PERK pathway. *Photodiagn. Photodyn. Ther.* **2021**, *34*, 102305. [CrossRef]
105. Velazquez, O.C. Angiogenesis and vasculogenesis: Inducing the growth of new blood vessels and wound healing by stimulation of bone marrow-derived progenitor cell mobilization and homing. *J. Vasc. Surg.* **2007**, *45* (Suppl. A), A39–A47. [CrossRef]
106. Mroz, P.; Yaroslavsky, A.; Kharkwal, G.B.; Hamblin, M.R. Cell death pathways in photodynamic therapy of cancer. *Cancers* **2011**, *3*, 2516–2539. [CrossRef] [PubMed]
107. Castano, A.P.; Demidova, T.N.; Hamblin, M.R. Mechanisms in photodynamic therapy: Part one-photosensitizers, photochemistry and cellular localization. *Photodiagn. Photodyn. Ther.* **2004**, *1*, 279–293. [CrossRef] [PubMed]
108. Verger, A.; Brandhonneur, N.; Molard, Y.; Cordier, S.; Kowouvi, K.; Amela-Cortes, M.; Dollo, G. From molecules to nanovectors: Current state of the art and applications of photosensitizers in photodynamic therapy. *Int. J. Pharm.* **2021**, *604*, 120763. [CrossRef]
109. Dobson, J.; de Queiroz, G.F.; Golding, J.P. Photodynamic therapy and diagnosis: Principles And comparative aspects. *Vet. J.* **2018**, *233*, 8–18. [CrossRef]
110. Mokoena, D.R.; George, B.P.; Abrahamse, H. Photodynamic Therapy Induced Cell Death Mechanisms in Breast Cancer. *Int. J. Mol. Sci.* **2021**, *22*, 10506. [CrossRef] [PubMed]
111. de Almeida, R.M.S.; Fontana, L.C.; Dos Santos Vitorio, G.; Pereira, A.H.C.; Soares, C.P.; Pinto, J.G.; Ferreira-Strixino, J. Analysis of the effect of photodynamic therapy with Fotoenticine on gliosarcoma cells. *Photodiagn. Photodyn. Ther.* **2020**, *30*, 101685. [CrossRef]
112. Rasmussen-Taxdal, D.S.; Ward, G.E.; Figge, F.H. Fluorescence of Human Lymphatic and Cancer Tissues Following High Doses of Intravenous Hematoporphyrin. *Surg. Forum* **1955**, *5*, 619–624. [CrossRef]
113. Kou, J.; Dou, D.; Yang, L. Porphyrin Photosensitizers in Photodynamic Therapy and its Applications. *Oncotarget* **2017**, *8*, 81591–81603. [CrossRef]
114. Gunaydin, G.; Gedik, M.E.; Ayan, S. Photodynamic Therapy for the Treatment and Diagnosis of Cancer-A Review of the Current Clinical Status. *Front. Chem.* **2021**, *9*, 686303. [CrossRef]
115. Agostinis, P.; Berg, A.; Cengel, K.A.; Foster, T.H.; Girotti, A.W.; Golinick, S.O.; Hahn, S.M.; Hamblin, M.R.; Juzeniene, A.; Kessel, D.; et al. Photodynamic therapy of cancer: An update. *CA Cancer* **2011**, *61*, 250–281. [CrossRef]
116. Dougherty, J.; Gomer, C.J.; Henderson, B.W.; Jori, G.; Kessel, D.; Korbelik, M.; Moan, J.; Peng, Q. Photodynamic therapy. *J. Natl. Cancer Inst.* **1998**, *90*, 889–905. [CrossRef] [PubMed]
117. Przygoda, M.; Bartusik-Aebisher, D.; Dynarowicz, K.; Cieślar, G.; Kawczyk-Krupka, A.; Aebisher, D. Cellular Mechanisms of Singlet Oxygen in Photodynamic Therapy. *Int. J. Mol. Sci.* **2023**, *24*, 16890. [CrossRef]
118. Dolmans, E.; Fukumura, D.; Jain, R.K. Photodynamic therapy for cancer. *Nat. Rev. Cancer* **2003**, *3*, 380–387. [CrossRef]
119. Lima, E.; Reis, L.V. Photodynamic Therapy: From the Basics to the Current Progress of N-Heterocyclic-Bearing Dyes as Effective Photosensitizers. *Molecules* **2023**, *28*, 5092. [CrossRef]
120. Phillips, D. Light relief: Photochemistry and medicine. *Photochem. Photobiol. Sci.* **2010**, *9*, 1589–1596. [CrossRef]
121. Thomas, E.; Colombeau, L.; Gries, M.; Peterlini, T.; Mathieu, C.; Thomas, N.; Boura, C.; Frochot, C.; Vanderesse, R.; Lux, F.; et al. Ultrasmall AGuIX theranostic nanoparticles for vascular-targeted interstitial photodynamic therapy of glioblastoma. *Int. J. Nanomed.* **2017**, *12*, 7075–7088. [CrossRef]
122. Sobhani, N.; Samadani, A.A. Implications of photodynamic cancer therapy: An overview of PDT mechanisms basically and practically. *J. Egypt. Natl. Canc Inst.* **2021**, *33*, 34. [CrossRef] [PubMed]
123. Bartusik-Aebisher, D.; Żołyniak, A.; Barnaś, E.; Machorowska-Pieniążek, A.; Oleś, P.; Kawczyk-Krupka, A.; Aebisher, D. The Use of Photodynamic Therapy in the Treatment of Brain Tumors-A Review of the Literature. *Molecules* **2022**, *27*, 6847. [CrossRef]
124. Eljamel, M.S. New light on the brain: The role of photosensitising agents and laser light in the management of invasive intracranial tumours. *Technol. Cancer Res. Treat.* **2003**, *2*, 303–309. [CrossRef] [PubMed]
125. Fisher, C.J.; Lilge, L. Photodynamic therapy in the treatment of intracranial gliomas: A review of current practice and considerations for future clinical directions. *J. Inn. Opt. Health Sci.* **2015**, *8*, 1530005. [CrossRef]
126. Hirschberg, H.; Berg, K.; Peng, Q. Photodynamic therapy mediated immune therapy of brain tumors. *Neuroimmunol. Neuroinflamm.* **2018**, *5*, 27. [CrossRef] [PubMed]
127. Huis In't Veld, R.V.; Heuts, J.; Ma, S.; Cruz, L.J.; Ossendorp, F.A.; Jager, M.J. Current Challenges and Opportunities of Photodynamic Therapy against Cancer. *Pharmaceutics* **2023**, *5*, 330. [CrossRef]
128. Ibarra, L.E.; Vilchez, M.L.; Caverzán, M.D.; Sanabria, L.M.N. Understanding the glioblastoma tumor biology to optimize photodynamic therapy: From molecular to cellular events. *J. Neursci. Res.* **2021**, *99*, 1024–1047. [CrossRef] [PubMed]
129. Shahmoradi Ghahe, S.; Kosicki, K.; Wojewódzka, M.; Majchrzak, B.A.; Fogtman, A.; Iwanicka-Nowicka, R.; Ciuba, A.; Koblowska, M.; Kruszewski, M.; Tudek, B.; et al. Increased DNA repair capacity augments resistance of glioblastoma cells to photodynamic therapy. *DNA Repair* **2021**, *104*, 103136. [CrossRef] [PubMed]
130. Ng, J.; Henriquez, N.; MacRobert, A.; Kitchen, N.; Williams, N.; Bown, S. Bioluminescence-activated photodynamic therapy for luciferase transfected, grade 4 astrocytoma cells in vitro. *Photodiagn. Photodyn. Ther.* **2022**, *38*, 102856. [CrossRef] [PubMed]
131. Xu, H.Z.; Li, T.F.; Ma, Y.; Li, K.; Zhang, Q.; Xu, Y.H.; Zhang, Y.C.; Zhao, L.; Chen, X. Targeted photodynamic therapy of glioblastoma mediated by platelets with photo-controlled release property. *Biomaterials* **2022**, *290*, 121833. [CrossRef]

132. Lv, Z.; Cao, Y.; Xue, D.; Zhang, H.; Zhou, S.; Yin, N.; Li, W.; Jin, L.; Wang, Y.; Zhang, H. A multiphoton transition activated iron based metal organic framework for synergistic therapy of photodynamic therapy/chemodynamic therapy/chemotherapy for orthotopic gliomas. *J. Mater. Chem. B* **2023**, *11*, 1100–1107. [CrossRef]
133. Christie, C.; Madsen, S.J.; Peng, Q.; Hirschberg, H. Macrophages as a photosensitizer delivery system for photodynamic therapy: Potential for the local treatment of resected glioblastoma. *Photodiagn. Photodyn. Ther.* **2023**, *45*, 103897. [CrossRef]
134. Balalaeva, I.V.; Mishchenko, T.A.; Turubanova, V.D.; Peskova, N.N.; Shilyagina, N.Y.; Plekhanov, V.I.; Lermontova, S.A.; Klapshina, L.G.; Vedunova, M.V.; Krysko, D.V. Cyanoarylporphyrazines with High Viscosity Sensitivity: A Step towards Dosimetry-Assisted Photodynamic Cancer Treatment. *Molecules* **2021**, *26*, 5816. [CrossRef] [PubMed]
135. Caverzán, M.D.; Beaugé, L.; Chesta, C.A.; Palacios, R.E.; Ibarra, L.E. Photodynamic therapy of Glioblastoma cells using doped conjugated polymer nanoparticles: An in vitro comparative study based on redox status. *J. Photochem. Photobiol. B* **2020**, *212*, 112045. [CrossRef]
136. van Straten, D.; Mashayekhi, V.; de Bruijn, H.S.; Oliveira, S.; Robinson, D.J. Oncologic Photodynamic Therapy: Basic Principles, Current Clinical Status and Future Directions. *Cancers* **2017**, *9*, 19. [CrossRef] [PubMed]
137. Farberg, A.S.; Marson, J.W.; Soleymani, T. Advances in Photodynamic Therapy for the Treatment of Actinic Keratosis and Nonmelanoma Skin Cancer: A Narrative Review. *Dermatol. Ther.* **2023**, *13*, 689–716. [CrossRef]
138. Alsaab, H.O.; Alghamdi, M.S.; Alotaibi, A.S.; Alzhrani, R.; Alwuthaynani, F.; Althobaiti, Y.S.; Almalki, A.H.; Sau, S.; Iyer, A.K. Progress in Clinical Trials of Photodynamic Therapy for Solid Tumors and the Role of Nanomedicine. *Cancers* **2020**, *12*, 2793. [CrossRef]
139. Barbosa, H.F.G.; Piva, H.L.; Matsuo, F.S.; de Lima, S.C.G.; de Souza, L.E.B.; Osako, M.K.; Tedesco, A.C. Hybrid lipid-biopolymer nanocarrier as a strategy for GBM photodynamic therapy (PDT). *Int. J. Biol. Macromol.* **2023**, *242 Pt 1*, 124647. [CrossRef]
140. Zhang, B.; Xue, R.; Sun, C. Rational design of ROS-responsive nanocarriers for targeted X-ray-induced photodynamic therapy and cascaded chemotherapy of intracranial glioblastoma. *Nanoscale* **2022**, *14*, 5054–5067. [CrossRef]
141. Arias-Ramos, N.; Ibarra, L.E.; Serrano-Torres, M.; Yagüe, B.; Caverzán, M.D.; Chesta, C.A.; Palacios, R.E.; López-Larrubia, P. Iron Oxide Incorporated Conjugated Polymer Nanoparticles for Simultaneous Use in Magnetic Resonance and Fluorescent Imaging of Brain Tumors. *Pharmaceutics* **2021**, *13*, 1258. [CrossRef]
142. Ibarra, L.E.; Porcal, G.V.; Macor, L.P.; Ponzio, R.A.; Spada, R.M.; Lorente, C.; Chesta, C.A.; Rivarola, V.A.; Palacios, R.E. Metallated porphyrin-doped conjugated polymer nanoparticles for efficient photodynamic therapy of brain and colorectal tumor cells. *Nanomedicine* **2018**, *13*, 605–624. [CrossRef]
143. Comincini, S.; Manai, F.; Sorrenti, M.; Perteghella, S.; D'Amato, C.; Miele, D.; Catenacci, L.; Bonferoni, M.C. Development of Berberine-Loaded Nanoparticles for Astrocytoma Cells Administration and Photodynamic Therapy Stimulation. *Pharmaceutics* **2023**, *15*, 1078. [CrossRef]
144. Liu, Y.; Dai, S.; Wen, L.; Zhu, Y.; Tan, Y.; Qiu, G.; Meng, T.; Yu, F.; Yuan, H.; Hu, F. Enhancing Drug Delivery for Overcoming Angiogenesis and Improving the Phototherapy Efficacy of Glioblastoma by ICG-Loaded Glycolipid-Like Micelles. *Int. J. Nanomed.* **2020**, *15*, 2717–2732. [CrossRef]
145. Teng, C.W.; Amirshaghaghi, A.; Cho, S.S.; Cai, S.S.; De Ravin, E.; Singh, Y.; Miller, J.; Sheikh, S.; Delikatny, E.; Cheng, Z.; et al. Combined fluorescence-guided surgery and photodynamic therapy for glioblastoma multiforme using cyanine and chlorin nanocluster. *J. Neurooncol.* **2020**, *149*, 243–252. [CrossRef]
146. Muhr, V.; Würth, C.; Kraft, M.; Buchner, M.; Baeumner, A.J.; Resch-Genger, U.; Hirsch, T. Particle-Size-Dependent Förster Resonance Energy Transfer from Upconversion Nanoparticles to Organic Dyes. *Anal. Chem.* **2017**, *89*, 4868–4874. [CrossRef]
147. Rai, P.; Mallidi, S.; Zheng, X.; Rahmanzadeh, R.; Mir, Y.; Elrington, S.; Khurshid, A.; Hasan, T. Development and Applications of Photo-Triggered Theranostic Agents. *Adv. Drug Deliv. Rev.* **2010**, *62*, 1094–1124. [CrossRef]
148. Guo, H.; Qian, H.; Idris, N.M.; Zhang, Y. Singlet Oxygen-Induced Apoptosis of Cancer Cells using Upconversion Fluorescent Nanoparticles as a Carrier of Photosensitizer. *Nanomed. Nanotechnol. Biol. Med.* **2010**, *6*, 486–495. [CrossRef] [PubMed]
149. Zhou, A.; Wei, Y.; Wu, B.; Chen, Q.; Xing, D. Pyropheophorbide A and c(RGDyK) Comodified Chitosan-Wrapped Upconversion Nanoparticle for Targeted Near-Infrared Photodynamic Therapy. *Mol. Pharm.* **2012**, *9*, 1580–1589. [CrossRef] [PubMed]
150. Idris, N.M.; Gnanasammandhan, M.K.; Zhang, J.; Ho, P.C.; Mahendran, R.; Zhang, Y. In Vivo Photodynamic Therapy using Upconversion Nanoparticles as Remote-Controlled Nanotransducers. *Nat. Med.* **2012**, *18*, 1580–1585. [CrossRef]
151. Cui, S.; Yin, D.; Chen, Y.; Di, Y.; Chen, H.; Ma, Y.; Achilefu, S.; Gu, Y. In Vivo Targeted Deep-Tissue Photodynamic Therapy Based on Near-Infrared Light Triggered Upconversion Nanoconstruct. *ACS Nano* **2012**, *7*, 676–688. [CrossRef] [PubMed]
152. Wang, C.; Cheng, L.; Liu, Z. Upconversion Nanoparticles for Photodynamic Therapy and other Cancer Therapeutics. *Theranostics* **2013**, *3*, 317–330. [CrossRef] [PubMed]
153. Wang, X.; Yang, C.-X.; Chen, J.-T.; Yan, X.-P. A Dual-Targeting Upconversion Nanoplatform for Two-Color Fluorescence Imaging-Guided Photodynamic Therapy. *Anal. Chem.* **2014**, *86*, 3263–3267. [CrossRef] [PubMed]
154. Lucky, S.S.; Idris, N.M.; Li, Z.; Huang, K.; Soo, K.C.; Zhang, Y. Titania Coated Upconversion Nanoparticles for Near-Infrared Light Triggered Photodynamic Therapy. *ACS Nano* **2015**, *9*, 191–205. [CrossRef] [PubMed]
155. Hou, Z.; Zhang, Y.; Deng, K.; Chen, Y.; Li, X.; Deng, X.; Cheng, Z.; Lian, H.; Li, C.; Lin, J. UV-Emitting Upconversion-Based TiO2 Photosensitizing Nanoplatform: Near-Infrared Light Mediated in Vivo Photodynamic Therapy via Mitochondria-Involved Apoptosis Pathway. *ACS Nano* **2015**, *9*, 2584–2599. [CrossRef] [PubMed]

156. Rai, M.; Singh, S.K.; Singh, A.K.; Prasad, R.; Koch, B.; Mishra, K.; Rai, S.B. Enhanced Red Upconversion Emission, Magneto-luminescent Behavior, and Bioimaging Application of $NaSc_{0.75}Er_{0.02}Yb_{0.18}Gd_{0.05}F_4$@ AuNPs Nanoparticles. *ACS Appl. Mater. Interfaces* **2015**, *7*, 15339–15350. [CrossRef] [PubMed]
157. Yang, G.; Yang, D.; Yang, P.; Lv, R.; Li, C.; Zhong, C.; He, F.; Gai, S.; Lin, J. A Single 808 nm Near-Infrared Light-Mediated Multiple Imaging and Photodynamic Therapy Based on Titania Coupled Upconversion Nanoparticles. *Chem. Mater.* **2015**, *27*, 7957–7968. [CrossRef]
158. Moan, J. Properties for optimal PDT sensitizers. *J. Photochem. Photobiol.* **1990**, *5*, 521–524. [CrossRef] [PubMed]
159. Ormond, A.B.; Freeman, H.S. Dye Sensitizers for Photodynamic Therapy. *Materials* **2013**, *6*, 817–840. [CrossRef]
160. Allison, R.R.; Downie, G.H.; Cuenca, R.; Hu, X.H.; Childs, C.J.; Sibata, C.H. Photosensitizers in clinical PDT. *Photodiagn. Photodyn. Ther.* **2004**, *1*, 27–42. [CrossRef]
161. Lange, C.; Bednarski, P.J. Photosensitizers for Photodynamic Therapy: Photochemistry in the Service of Oncology. *Curr. Pharm. Des.* **2016**, *22*, 6956–6974. [CrossRef]
162. Lan, M.; Zhao, S.; Liu, W.; Lee, C.S.; Zhang, W.; Wang, P. Photosensitizers for Photodynamic Therapy. *Adv. Healthc. Mater.* **2019**, *8*, e1900132. [CrossRef] [PubMed]
163. Kubrak, T.P.; Kołodziej, P.; Sawicki, J.; Mazur, A.; Koziorowska, K.; Aebisher, D. Some Natural Photosensitizers and Their Medicinal Properties for Use in Photodynamic Therapy. *Molecules* **2022**, *27*, 1192. [CrossRef] [PubMed]
164. Du, P.; Hu, S.; Cheng, Y.; Li, F.; Li, M.; Li, J.; Yi, L.; Feng, H. Photodynamic Therapy Leads to Death of C6 Glioma Cells Partly through AMPAR. *Brain Res.* **2012**, *1433*, 153–159. [CrossRef]
165. Udrea, A.M.; Smarandache, A.; Dinache, A.; Mares, C.; Nistorescu, S.; Avram, S.; Staicu, A. Photosensitizers-Loaded Nanocarriers for Enhancement of Photodynamic Therapy in Melanoma Treatment. *Pharmaceutics* **2023**, *15*, 2124. [CrossRef]
166. Dąbrowski, J.M.; Arnaut, L.G. Photodynamic Therapy (PDT) of Cancer: From Local to Systemic Treatment. *Photochem. Photobiol. Sci.* **2015**, *14*, 1765–1780. [CrossRef]
167. Correia, J.H.; Rodrigues, J.A.; Pimenta, S.; Dong, T.; Yang, Z. Photodynamic Therapy Review: Principles, Photosensitizers. *Appl. Future Direct. Pharm.* **2021**, *13*, 1332.
168. Simões, J.C.S.; Sarpaki, S.; Papadimitroulas, P.; Therrien, B.; Loudos, G. Conjugated photosensitizers for imaging and PDT in cancer research. *J. Med. Chem.* **2020**, *63*, 14119–14150. [CrossRef]
169. Abrahamse, H.; Hamblin, M.R. New photosensitizers for photodynamic therapy. *Biochem. J.* **2016**, *473*, 347–364. [CrossRef]
170. Kwiatkowski, S.; Knap, B.; Przystupski, D.; Saczko, J.; Kędzierska, E.; Knap-Czop, K.; Kotlińska, J.; Michel, O.; Kotowski, K.; Kulbacka, J. Photodynamic therapy—Mechanisms, photosensitizers and combinations. *Biomed. Pharmacother.* **2018**, *106*, 1098–1107. [CrossRef] [PubMed]
171. Mahmoudi, K.; Garvey, K.L.; Bouras, A.; Cramer, G.; Stepp, H.; Raj, J.G.J.; Bozec, D.; Busch, T.; Hadjipanayis, C.G. 5-aminolevulinic acid photodynamic therapy for the treatment of high-grade gliomas. *J. Neuro-Oncol.* **2019**, *141*, 595–607. [CrossRef] [PubMed]
172. Wachowska, M.; Muchowicz, A.; Firczuk, M.; Gabrysiak, M.; Winiarska, M.; Wańczyk, M.; Bojarczuk, K.; Golab, J. Aminolevulinic Acid (ALA) as a Prodrug in Photodynamic Therapy of Cancer. *Molecules* **2011**, *16*, 4140–4164. [CrossRef]
173. Yang, X.; Palasuberniam, P.; Kraus, D.; Chen, B. Aminolevulinic Acid-Based Tumor Detection and Therapy: Molecular Mechanisms and Strategies for Enhancement. *Int. J. Mol. Sci.* **2015**, *16*, 25865–25880. [CrossRef]
174. Pedrosa, L.; Bedia, C.; Diao, D.; Mosteiro, A.; Ferrés, A.; Stanzani, E.; Martínez-Soler, F.; Tortosa, A.; Pineda, E.; Aldecoa, I.; et al. Preclinical Studies with Glioblastoma Brain Organoid Co-Cultures Show Efficient 5-ALA Photodynamic Therapy. *Cells* **2023**, *12*, 1125. [CrossRef]
175. Omura, N.; Nonoguchi, N.; Fujishiro, T.; Park, Y.; Ikeda, N.; Kajimoto, Y.; Hosomi, R.; Yagi, R.; Hiramatsu, R.; Furuse, M.; et al. Ablation efficacy of 5-aminolevulinic acid-mediated photodynamic therapy on human glioma stem cells. *Photodiagnosis Photodyn. Ther.* **2023**, *41*, 103119. [CrossRef]
176. Casas, A. Clinical uses of 5-aminolaevulinic acid in photodynamic treatment and photodetection of cancer: A review. *Cancer Lett.* **2020**, *490*, 165–173. [CrossRef]
177. Müller, P.; Abdel Gaber, S.A.; Zimmermann, W.; Wittig, R.; Stepp, H. ABCG2 influence on the efficiency of photodynamic therapy in glioblastoma cells. *J. Photochem. Photobiol. B* **2020**, *210*, 111963. [CrossRef]
178. Stepp, H.; Stummer, W. 5-ALA in the management of malignant glioma. *Lasers Surg. Med.* **2018**, *50*, 399–419. [CrossRef]
179. Mansi, M.; Howley, R.; Chandratre, S.; Chen, B. Inhibition of ABCG2 transporter by lapatinib enhances 5-aminolevulinic acid-mediated protoporphyrin IX fluorescence and photodynamic therapy response in human glioma cell lines. *Biochem. Pharmacol.* **2022**, *200*, 115031. [CrossRef]
180. Kast, R.E.; Michael, A.P.; Sardi, I.; Burns, T.C.; Heiland, T.; Karpel-Massler, G.; Kamar, F.G.; Halatsch, M.E. A New Treatment Opportunity for DIPG and Diffuse Midline Gliomas: 5-ALA Augmented Irradiation, the 5aai Regimen. *Brain Sci.* **2020**, *10*, 51. [CrossRef] [PubMed]
181. Schipmann, S.; Müther, M.; Stögbauer, L.; Zimmer, S.; Brokinkel, B.; Holling, M.; Grauer, O.; Suero Molina, E.; Warneke, N.; Stummer, W. Combination of ALA-induced fluorescence-guided resection and intraoperative open photodynamic therapy for recurrent glioblastoma: Case series on a promising dual strategy for local tumor control. *J. Neurosurg.* **2020**, *134*, 426–436. [CrossRef]
182. McNicholas, K.; MacGregor, M.; Gleadle, J. In order for the light to shine so brightly, the darkness must be present—Why do cancers fluoresce with 5-aminolaevulinic acid? *Br. J. Cancer* **2019**, *121*, 631–639. [CrossRef] [PubMed]

183. Mazurek, M.; Szczepanek, D.; Orzyłowska, A.; Rola, R. Analysis of Factors Affecting 5-ALA Fluorescence Intensity in Visualizing Glial Tumor Cells—Literature Review. *Int. J. Mol. Sci.* **2022**, *23*, 926. [CrossRef]
184. Ihata, T.; Nonoguchi, N.; Fujishiro, T.; Omura, N.; Kawabata, S.; Kajimoto, Y.; Wanibuchi, M. The effect of hypoxia on photodynamic therapy with 5-aminolevulinic acid in malignant gliomas. *Photodiagn. Photodyn. Ther.* **2022**, *40*, 103056. [CrossRef] [PubMed]
185. Leroy, H.A.; Guérin, L.; Lecomte, F.; Baert, G.; Vignion, A.S.; Mordon, S.; Reyns, N. Is interstitial photodynamic therapy for brain tumors ready for clinical practice? A systematic review. *Photodiagn. Photodyn. Ther.* **2021**, *36*, 102492. [CrossRef] [PubMed]
186. Kimura, S.; Kuroiwa, T.; Ikeda, N.; Nonoguchi, N.; Kawabata, S.; Kajimoto, Y.; Ishikawa, T. Assessment of safety of 5-aminolevulinic acid-mediated photodynamic therapy in rat brain. *Photodiagn. Photodyn. Ther.* **2018**, *21*, 367–374. [CrossRef] [PubMed]
187. Perria, C.; Capuzzo, T.; Cavagnaro, G.; Datti, R.; Francaviglia, N.; Rivano, C.; Tercero, V.E. First attempts at the photodynamic treatment of human gliomas. *J. Neurosurg. Sci.* **1980**, *24*, 119–129.
188. Kostron, H.; Hochleitner, B.W.; Obwegeser, A.; Seiwald, M. Clinical and experimental results of photodynamic therapy in neurosurgery. *SPIE Proc.* **1995**, *2371*, 126–128.
189. Stylli, S.S.; Howes, M.; MacGregor, L.; Rajendra, P.; Kaye, A.H. Photodynamic therapy of brain tumours: Evaluation of porphyrin uptake versus clinical outcome. *J. Clin. Neurosci.* **2004**, *11*, 584–596. [CrossRef]
190. Kaye, A.H.; Hill, J.S. Photodynamic therapy of brain tumours. *Ann. Acad. Med. Singap.* **1993**, *22*, 470–481.
191. Laws Jr, E.R.; Cortese, D.A.; Kinsey, J.H.; Eagan, R.T.; Anderson, R.E. Photoradiation therapy in the treatment of malignant brain tumors: A phase I (feasibility) study. *Neurosurgery* **1981**, *9*, 672–678. [CrossRef]
192. McCulloch, G.A.J.; Forbes, I.J.; Lee See, K. Phototherapy in malignant brain tumours. In *Porphyrin Localization and Treatment of Tumours*; Doiron, D.R., Gomer, C.J., Eds.; Alan R. Liss: New York, NY, USA, 1984; pp. 709–717.
193. Kaye, A.H.; Morstyn, G.; Brownbill, D. Adjuvant high-dose photoradiation therapy in the treatment of cerebral glioma: A phase 1-2 study. *J. Neurosurg.* **1987**, *67*, 500–505. [CrossRef] [PubMed]
194. Vilchez, M.L.; Rodríguez, L.B.; Palacios, R.E.; Prucca, C.G.; Caverzán, M.D.; Caputto, B.L.; Rivarola, V.A.; Milla Sanabria, L.N. Isolation and initial characterization of human glioblastoma cells resistant to photodynamic therapy. *Photodiagn. Photodyn. Ther.* **2021**, *33*, 102097. [CrossRef] [PubMed]
195. Quirka, B.J.; Brandal, G.; Donlon, S.; Vera, J.C.; Mang, T.S.; Foy, A.B.; Lew, S.M.; Girotti, A.W.; Jogal, S.; LaViolette, P.S.; et al. Photodynamic therapy (PDT) for malignant brain tumors-Where do we stand? *Photodiagn. Photodyn. Ther.* **2015**, *12*, 530–544. [CrossRef] [PubMed]
196. Kaneko, S.; Fujimoto, S.; Yamaguchi, H.; Yamauchi, T.; Yoshimoto, T.; Tokuda, K. Photodynamic therapy of malignant gliomas. In *Intracranial Gliomas Part III-Innovative Treatment Modalities*; Progress in Neurological Surgery; Chernov, M.F., Muragaki, Y., Kesari, S., McCutcheon, I.E., Eds.; Karger: Basel, Switzerland, 2018; Volume 32, pp. 1–13.
197. Ren, Z.; Wen, J.; Mo, Y.; Zhang, P.; Chen, H.; Wen, J. A systematic review and meta-analysis offluorescent-guided resection and therapy-based photodynamics on the survival of patients with glioma. *Lasers Med. Sci.* **2022**, *37*, 789–797. [CrossRef] [PubMed]
198. Muller, P.; Wilson, B. A randomized two arm clinical trial of photophrin PDT and standard therapy in high grade gliomas—Phase III trial. In Proceedings of the 6th International Symposium on Photodynamic Diagnosis and Therapy in Clinical Practice, Brixen/Bressanone, Italy, 10–14 October 2006.
199. Kostron, H.; Fiegele, T.; Akatuna, E. Combination of FOSCAN® mediated fluorescence guided resection and photodynamic treatment as new therapeutic concept for malignant brain tumors. *Med. Laser Appl.* **2006**, *21*, 285–290. [CrossRef]
200. Zhan, Q.; Yue, W.; Shaoshan, H. The inhibitory effect of photodynamic therapy and of an anti-VCAM-1 monoclonal antibody on the in vivo growth of C6 glioma xenografts. *Braz. J. Med. Biol. Res.* **2011**, *44*, 489–490. [CrossRef] [PubMed]
201. Muragaki, Y.; Akimoto, J.; Maruyama, T.; Iseki, H.; Ikuta, S.; Nitta, M.; Maebayashi, K.; Saito, T.; Okada, Y.; Kaneko, S.; et al. Phase II clinical study on intraoperative photodynamic therapy with talaporfin sodium and semiconductor laser in patients with malignant brain tumors. *J. Neurosurg.* **2013**, *119*, 845–852. [CrossRef] [PubMed]
202. Eljamel, M.S.; Goodman, C.; Moseley, H. ALA and Photofrin fluorescence-guided resection and repetitive PDT in glioblastoma multiforme: A single centre Phase III randomized controlled trial. *Lasers Med. Sci.* **2008**, *23*, 361–367. [CrossRef] [PubMed]
203. Turubanova, V.D.; Balalaeva, I.V.; Mishchenko, T.A.; Catanzaro, E.; Alzeibak, R.; Peskova, N.N.; Efimova, I.; Bachert, C.; Mitroshina, E.V.; Krysko, O.; et al. Immunogenic cell death induced by a new photodynamic therapy based on photosens and photodithazine. *J. Immunother. Cancer* **2019**, *7*, 350. [CrossRef]
204. Hu, S.L.; Du, P.; Hu, R.; Li, F.; Feng, H. Imbalance of Ca^{2+} and K^{+} fluxes in C6 glioma cells after PDT measured with scanning ion-selective electrode technique. *Lasers Med. Sci.* **2014**, *29*, 1261–1267. [CrossRef]
205. Wu, D.P.; Bai, L.R.; Lv, Y.F.; Zhou, Y.; Ding, C.H.; Yang, S.M.; Zhang, F.; Huang, J.L. A novel role of Cx43-composed GJIC in PDT phototoxicity: An implication of Cx43 for the enhancement of PDT efficacy. *Int. J. Biol. Sci.* **2019**, *15*, 598–609. [CrossRef]
206. Kammerer, R.; Buchner, A.; Palluch, P.; Pongratz, T.; Oboukhovskij, K.; Beyer, W.; Johansson, A.; Stepp, H.; Baumgartner, R.; Zimmermann, W. Induction of immune mediators in glioma and prostate cancer cells by non-lethal photodynamic therapy. *PLoS ONE* **2011**, *6*, e21834. [CrossRef]
207. Fisher, C.J.; Niu, C.; Foltz, W.; Chen, Y.; Sidorova-Darmos, E.; Eubanks, J.H.; Lilge, L. ALA-PpIX mediated photodynamic therapy of malignant gliomas augmented by hypothermia. *PLoS ONE* **2017**, *12*, e0181654. [CrossRef]

208. Zhang, F.; Wen, C.; Peng, Y.; Hu, Z.; Zheng, S.; Chen, W.; Wen, L. Biomimetic lipid nanoparticles for homologous-targeting and enhanced photodynamic therapy against glioma. *Eur. J. Pharm. Sci.* **2023**, *190*, 106574. [CrossRef]
209. Jiang, F.; Robin, A.M.; Katakowski, M.; Tong, L.; Espiritu, M.; Singh, G.; Chopp, M. Photodynamic therapy with photofrin in combination with Buthionine Sulfoximine (BSO) of human glioma in the nude rat. *Lasers Med. Sci.* **2003**, *18*, 128–133. [CrossRef]
210. Yi, W.; Xu, H.T.; Tian, D.F.; Wu, L.Q.; Zhang, S.Q.; Wang, L.; Ji, B.W.; Zhu, X.N.; Okechi, H.; Liu, G.; et al. Photodynamic therapy mediated by 5-aminolevulinic acid suppresses gliomas growth by decreasing the microvessels. *J. Huazhong. Univ. Sci. Technol. Med. Sci.* **2015**, *35*, 259–264. [CrossRef]
211. Zheng, X.; Chopp, M.; Lu, Y.; Jiang, J.; Zhao, D.; Ding, C.; Yang, H.; Zhang, L.; Jiang, F. Atorvastatin reduces functional deficits caused by photodynamic therapy in rats. *Int. J. Oncol.* **2011**, *39*, 1133–1141.
212. Hou, K.; Liu, J.; Du, J.; Mi, S.; Ma, S.; Ba, Y.; Ji, H.; Li, B.; Hu, S. Dihydroartemisinin prompts amplification of photodynamic therapy-induced reactive oxygen species to exhaust Na/H exchanger 1-mediated glioma cells invasion and migration. *J. Photochem. Photobiol. B* **2021**, *219*, 112192. [CrossRef]
213. Terzis, A.J.; Dietze, A.; Bjerkvig, R.; Arnold, H. Effects of photodynamic therapy on glioma spheroids. *Br. J. Neurosurg.* **1997**, *11*, 196–205. [PubMed]
214. Benayoun, L.; Schaffer, M.; Bril, R.; Gingis-Velitski, S.; Segal, E.; Nevelsky, A.; Satchi-Fainaro, R.; Shaked, Y. Porfimer-sodium (Photofrin-II) in combination with ionizing radiation inhibits tumor-initiating cell proliferation and improves glioblastoma treatment efficacy. *Cancer Biol Ther.* **2013**, *14*, 64–74. [CrossRef]
215. Gao, S.-G.; Wang, L.-D.; Feng, X.-S.; Qu, Z.-F.; Shan, T.-Y.; Xie, X.-H. Absorption and elimination of photofrin-II in human immortalization esophageal epithelial cell line SHEE and its malignant transformation cell line SHEEC. *Ai Zheng* **2009**, *28*, 1248–1254. [CrossRef]
216. Fingar, V.H.; Wieman, T.J.; Haydon, P.S. The Effects of Thrombocytopenia on Vessel Stasis and Macromolecular Leakage after Photodynamic Therapy Using Photofrin. *Photochem. Photobiol.* **1997**, *66*, 513–517. [CrossRef] [PubMed]
217. Schmidt, M.H.; Meyer, G.A.; Reichert, K.W.; Cheng, J.; Krouwer, H.G.; Ozker, K.; Whelan, H.T. Evaluation of photodynamic therapy near functional brain tissue in patients with recurrent brain tumors. *J. Neurooncol.* **2004**, *67*, 201–207. [CrossRef] [PubMed]
218. Cramer, S.W.; Chen, C.C. Photodynamic Therapy for the Treatment of Glioblastoma. *Front. Surg.* **2019**, *6*, 81. [CrossRef] [PubMed]
219. Sharman, W.M.; Allen, C.M.; van Lier, J.E. Photodynamic Therapeutics: Basic Principles and Clinical Applications. *Drug Discov. Today* **1999**, *4*, 507–517. [CrossRef]
220. Schweitzer, V.G. Photodynamic Therapy for Treatment of Head and Neck Cancer. Otolaryngol. *Head Neck Surg.* **1990**, *102*, 225–232. [CrossRef] [PubMed]
221. Shimizu, K.; Nitta, M.; Komori, T.; Maruyama, T.; Yasuda, T.; Fujii, Y.; Masamune, K.; Kawamata, T.; Maehara, T.; Muragaki, Y. Intraoperative Photodynamic Diagnosis Using Talaporfin Sodium Simultaneously Applied for Photodynamic Therapy against Malignant Glioma: A Prospective Clinical Study. *Front Neurol.* **2018**, *9*, 24. [CrossRef] [PubMed]
222. Kim, M.M.; Darafsheh, A. Light Sources and Dosimetry Techniques for Photodynamic Therapy. *Photochem. Photobiol.* **2020**, *96*, 280–294. [CrossRef]
223. Kostron, H.; Obwegeser, A.; Jakober, R.; Zimmermann, A.; Rueck, A.C. Experimental and clinical results of mTHPC (Foscan)-mediated photodynamic therapy for malignant brain tumors. In *Optical Methods for Tumor Treatment and Detections: Mechanisms and Techniques in Photodynamic Therapy VII*; SPIE: Bellingham, WA, USA, 1998.
224. Callahan, D.E.; Forte, T.M.; Afzal, S.M.; Deen, D.F.; Kahl, S.B.; Bjornstad, K.A.; Bauer, W.F.; Blakely, E.A. Boronated Protoporphyrin (BOPP): Localization in Lysosomes of the Human Glioma Cell Line SF-767 with Uptake Modulated by Lipoprotein Levels. *Int. J. Radiat. Oncol. Biol. Phys.* **1999**, *45*, 761–771. [CrossRef]
225. Fingar, V.H.; Kik, P.K.; Haydon, P.S.; Cerrito, P.B.; Tseng, M.; Abang, E.; Wieman, T.J. Analysis of Acute Vascular Damage after Photodynamic Therapy Using Benzoporphyrin Derivative (BPD). *Br. J. Cancer* **1999**, *79*, 1702–1708. [CrossRef]
226. Aveline, B.; Hasan, T.; Redmond, R.W. Photophysical and Photosensitizing Properties of Benzoporphyrin Derivative Monoacid Ring A (BPD-MA). *Photochem. Photobiol.* **1994**, *59*, 328–335. [CrossRef]
227. Guo, H.W.; Lin, L.T.; Chen, P.H.; Ho, M.H.; Huang, W.T.; Lee, Y.J.; Chiou, S.H.; Hsieh, Y.S.; Dong, C.Y.; Wang, H.W. Low-fluence rate, long duration photodynamic therapy in glioma mouse model using organic light emitting diode (OLED). *Photodiagn. Photodyn. Ther.* **2015**, *12*, 504–510. [CrossRef]
228. Zhang, X.; Jiang, F.; Kalkanis, S.N.; Zhang, Z.; Hong, X.; Yang, H.; Chopp, M. Post-acute response of 9L gliosarcoma to Photofrin-mediated PDT in athymic nude mice. *Lasers Med. Sci.* **2007**, *22*, 253–259. [CrossRef]
229. ContributeCaverzán, M.D.; Oliveda, P.M.; Beaugé, L.; Palacios, R.E.; Chesta, C.A.; Ibarra, L.E. Metronomic Photodynamic Therapy with Conjugated Polymer Nanoparticles in Glioblastoma Tumor Microenvironment. *Cells* **2023**, *12*, 1541.
230. Ibarra, L.E.; Beaugé, L.; Arias-Ramos, N.; Rivarola, V.A.; Chesta, C.A.; Lopez-Larrubia, P.; Palacios, R.E. Trojan horse monocyte-mediated delivery of conjugated polymer nanoparticles for improved. *Nanomedicine* **2020**, *15*, 1687–1707. [CrossRef] [PubMed]
231. Vermandel, M.; Dupont, C.; Lecomte, F.; Leroy, H.A.; Tuleasca, C.; Mordon, S.; Hadjipanayis, C.G.; Reyns, N. Standardized intraoperative 5-ALA photodynamic therapy for newly diagnosed glioblastoma patients: A preliminary analysis of the INDYGO clinical trial. *J. Neurooncol.* **2021**, *152*, 501–514. [CrossRef] [PubMed]
232. Liu, Z.; Mela, A.; Argenziano, M.G.; Banu, M.A.; Furnari, J.; Kotidis, C.; Sperring, C.P.; Humala, N.; Mahajan, A.; Bruce, J.N.; et al. Single-cell analysis of 5-aminolevulinic acid intraoperative labeling specificity for glioblastoma. *J. Neurosurg.* **2023**, 1–11, *publish before print*. [CrossRef] [PubMed]

233. Whelan, H.; Strivatsal, S. *The Proverbial Light at the End of the Tunnel in Brain-Tumor Treatment*; SPIE: Bellingham, WA, USA, 2009; Volume 1670. [CrossRef]
234. Borgia, F.; Giuffrida, R.; Caradonna, E.; Vaccaro, M.; Guarneri, F.; Cannavò, S.P. Early and Late Onset Side Effects of Photodynamic Therapy. *Biomedicines* **2018**, *6*, 12. [CrossRef] [PubMed]
235. Kostron, H.; Weiser, G.; Fritsch, E.; Grunert, V. Photodynamic therapy of malignant brain tumors: Clinical and neuropathological results. *Photochem. Photobiol.* **1987**, *46*, 937–943. [CrossRef]
236. Stylli, S.S.; Kaye, A.H.; MacGregor, L.; Howes, M.; Rajendra, P. Photodynamic therapy of high grade glioma—Long term survival. *J. Clin. Neurosci.* **2005**, *12*, 389–398. [CrossRef]

biomedicines

MDPI

Review

Photodynamic Therapy for the Treatment of Bowen's Disease: A Review on Efficacy, Non-Invasive Treatment Monitoring, Tolerability, and Cosmetic Outcome

Paolo Antonetti [1,2], Cristina Pellegrini [1,2], Chiara Caponio [2], Manfredo Bruni [1,2], Lorenzo Dragone [1,2], Mirco Mastrangelo [1], Maria Esposito [1,2] and Maria Concetta Fargnoli [1,2,*]

1 Department of Biotechnological and Applied Clinical Sciences, University of L'Aquila, 67100 L'Aquila, Italy; paolo.antonetti@studenti.unich.it (P.A.); cristina.pellegrini@univaq.it (C.P.); manfredo.bruni@gmail.com (M.B.); mirco.mastrangelo@graduate.univaq.it (M.M.); maria.esposito3@univaq.it (M.E.)

2 Dermatology Unit, Ospedale San Salvatore, 67100 L'Aquila, Italy; chiara_caponio@hotmail.it

* Correspondence: mariaconcetta.fargnoli@univaq.it; Tel.: +39-0862-368519; Fax: +39-0862-433433

Abstract: Bowen's disease represents the in situ form of cutaneous squamous cell carcinoma; although it has an excellent prognosis, 3–5% of lesions progress to invasive cutaneous squamous cell carcinoma, with a higher risk in immunocompromised patients. Treatment is therefore always necessary, and conventional photodynamic therapy is a first-line option. The aim of this review is to provide an overview of the clinical response, recurrence rates, safety, and cosmetic outcome of photodynamic therapy in the treatment of Bowen's disease, considering different protocols in terms of photosensitizers, light source, and combination treatments. Photodynamic therapy is a valuable option for tumors at sites where wound healing is poor/delayed, in the case of multiple and/or large tumors, and where surgery would be difficult or invasive. Dermoscopy and reflectance confocal microscopy can be used as valuable tools for monitoring the therapeutic response. The treatment is generally well tolerated, with mild side effects, and is associated with a good/excellent cosmetic outcome. Periodic follow-up after photodynamic therapy is essential because of the risk of recurrence and progression to cSCC. As the incidence of keratinocyte tumors increases, the therapeutic space for photodynamic therapy will further increase.

Keywords: photodynamic therapy; Bowen's disease; squamous cell carcinoma in situ; MAL-PDT; ALA-PDT; non-melanoma skin cancer; immunocompromised patients

1. Introduction

Bowen's disease (BD) is the intraepidermal (in situ) form of cutaneous squamous cell carcinoma (cSCC), first described by Bowen in 1912 [1,2]. An average annual incidence of 22.4 lesions/100,000 women and 27.8/100,000 men has been reported in the 5-year period from 1996 to 2000 in Canada [3], and of 142 lesions/100,000 Caucasian residents from 1983–1987 in Hawai [4]. The standardized incidence ratio for in situ carcinoma of the skin is 65 times higher in renal transplant recipients than in the general population [5]. Unlike cSCC, BD appears to have a slight prevalence in women [6]. Risk factors for BD are Fitzpatrick phototype I-II, age over 60 years, chronic UV exposure, and immunosuppression, similar to invasive cSCC [7,8]. Other recognized risk factors include arsenic exposure [9,10] and HPV infections [11].

The classical clinical presentation of BD is an erythematous, scaly, slow-growing, well-demarcated patch or plaque, usually asymptomatic, although larger lesions may be associated with itching (Figure 1A). The overlying scales may be white or yellow and can be adherent or easily removed [12]. Less common variants include pigmented, subungual, periungual, palmar, perianal, and genital cSCC in situ, the latter referred to as Erythroplasia of Queyrat when it involves the penis. Regarding body sites, the most common anatomical sites for classic BD are the head and neck, followed by the limbs. The cheeks and lower limbs are more likely to be affected in women, while the bald scalp and ears are more often involved in men [13–15].

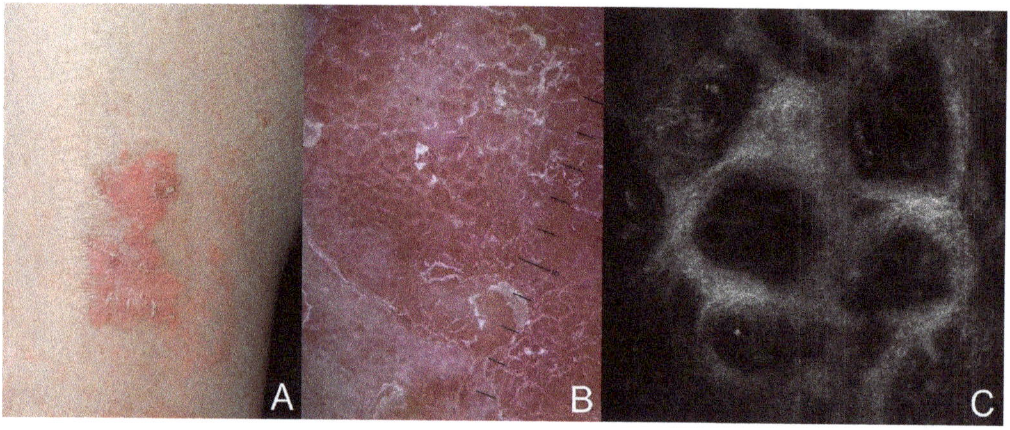

Figure 1. Clinical, dermoscopic, and confocal images of Bowen's disease: (**A**) erythematous, scaly, well-demarcated plaque; (**B**) glomerular vessels and scaly surface on an erythematous base (10×); and (**C**) tightly coiled vessels, some with an S-shape, in the center of dermal papillae; hyper-reflective stroma (mosaic, 8 × 8 mm).

Dermoscopy can be used as quick and non-invasive diagnostic technique for BD. The presence of vascular structures, i.e., dotted or glomerular vessels, a scaly surface, and white structureless areas characterize BD on dermoscopy (Figure 1B) [16–18]. In the pigmented variant, additional dermoscopic findings include small brown globules regularly packed in a patchy distribution, reticular pigmentation and structureless gray to brown pigmentation. Additional non-invasive imaging tools such as reflectance confocal microscopy (RCM) (Figure 1C), conventional optical coherence tomography (OCT), and line-field confocal optical coherence tomography (LC-OCT) can help in the early diagnosis of keratinocyte skin tumors [19–21].

Histologically, as a carcinoma in situ, BD shows full-thickness epidermal involvement but does not extend beyond the basal membrane. It is characterized by atypical keratinocytes, sometimes multinucleated, associated with a disordered maturation of the epidermis, dyskeratotic cells, and mitosis at various levels. A loss of the granular layer is usually present, with overlying parakeratosis and sometimes hyperkeratosis. An involvement of the pilosebaceous apparatus is possible [3,22].

Keratinocyte carcinomas are well known to occur at a higher rate in immunocompromised patients, including organ transplant recipients (OTRs) as well as patients with other forms of immune suppression (chronic leukemias, infections, and autoimmune diseases). Much of the existing literature derives from studies on OTRs, which represent the majority of the immunocompromised population. The cumulative incidence is related to geographic latitude, skin type, and immunosuppressive therapies [23]. Indeed, the risk appears to be correlated with the level of immunosuppression in the transplant (heart > kidney > liver) [24]. A multicenter US retrospective cohort study including 10,649 adults receiving a primary organ transplant in 2003 or 2008 reported that 8% developed skin cancer, yielding an incidence ratio of 1437 per 100,000 person-years, and 94% of them were cSCC, yielding an incidence ratio of 1355 per 100,000 person-years [25]. Detailed data are limited on BD lesions. In a large Irish population-based study in renal transplant recipients, BD lesions represented 19% of all cancer types with a 65-fold increased standardized incidence ratio (SIR 64.6; 95% CI 53.7–75.5) [5].

The prognosis of BD is excellent, as it is usually a slow-growing lesion. However, the overall rate of progression to invasive cSCC is 3–5%, or even up to 10% for genital lesions [3,6,15,26], and is more common among the elderly and immunocompromised individuals [27,28]. Clinical signs suggestive of malignant transformation are ulceration, nodule formation, and bleeding [6].

Treatment options for BD are multiple, including surgical excision, cryotherapy, laser ablation, curettage with cautery, radiation therapy, topical 5% 5-fluorouracil (5-FU), imiquimod, and conventional photodynamic therapy (PDT). Factors to consider when choosing treatment include the number, site, size, and thickness of the lesions, as well as comorbidities, immune status, and patient's preference [6,12,26]. Surgical excision is the first choice for the treatment of BD; however, non-invasive treatments are recognized as acceptable treatment options, with the opportunity to treat multiple lesions and the advantages of better cosmetic results and lower costs.

We performed a literature review on the application of PDT in the treatment of BD using the PubMed database, and the search terms were the following: photodynamic therapy, PDT, MAL-PDT, ALA-PDT, Bowen's disease, and squamous cell carcinoma in situ. This review includes studies published through January 2024, describing clinical response, recurrence rates, cosmetic outcome, tolerability, and the adverse effects of PDT in the treatment of BD, considering different protocols in terms of photosensitizers, light source, and combination treatments.

2. PDT: Mechanism of Action and Treatment Protocol

PDT consists of the activation of a photosensitizing drug by visible light. The sensitizer, when irradiated, generates reactive oxygen species, such as singlet oxygen, the hydroxyl radical, the superoxide anion and hydrogen peroxide, which have a direct cytotoxic effect, and stimulate the release of immune mediators, resulting in additional pro-inflammatory effects [29,30]. In dermatologic indications, PDT uses precursors of the heme biosynthetic pathway, particularly 5-aminolaevulinic acid (5-ALA) or its ester, methyl aminolaevulinate (MAL), as photosensitizers, which are converted within target cells into protoporphyrin IX (PpIX) and activated through a light source.

Among the agents licensed for PDT in Europe, MAL (160 mg/g, Metvix®/Metvixia®, Galderma, Paris, France) is the only one authorized for use together with red light to treat cSCC in situ/BD. No formulation of ALA-PDT is licensed for this indication.

Current European guidelines recommend MAL-PDT for the treatment of BD with a strength of recommendation A and quality of evidence 1 [31]. PDT is particularly indicated for lesions at sites of poor healing, for large or multiple lesions, and in cases where surgery would be difficult or invasive such as facial, digital, nail bed, and penile lesions. The protocol involves two MAL-PDT sessions 7 days apart, repeated at 3 months, if necessary. It is advisable to prepare treatment sites by the gentle removal of overlying scales and crusts with saline-soaked gauze or a curette/scalpel. MAL is applied for 3 h under occlusion, and the treatment sites are then illuminated with an appropriate light source. The light source used is red light (630 nm), which has a greater ability to penetrate tissue than green or blue light, at a dose of 37 J/cm^2 [31].

3. Efficacy of PDT in Monotherapy

The first large pan European study on MAL-PDT included 225 patients with histologically confirmed BD, randomized to MAL-PDT, cryotherapy, or topical 5% 5-FU for 4 weeks. After 3 months, lesion response rates were similar with all regimens (93% MAL-PDT, 86% cryotherapy, 83% 5-FU). At 12 months, the estimated sustained complete response rate with MAL-PDT was significantly higher than that with cryotherapy (80% vs. 67%, $p = 0.047$) and better than that with 5-FU (80% vs. 69%, $p = 0.19$). Maximum lesion diameter influenced the recurrence rate at 12 months after MAL-PDT, which was 10% in lesions up to 14 mm, 12%

in lesions 15 to 29 mm, and 30% in lesions 30 mm or larger. However, response appeared to be independent of lesion location [32].

Overall, the clinical response rate for MAL-PDT in the treatment of BD varies from 88–100% after one or two cycles at 3 months, with 68–89% of lesions clear over follow-up periods of 17–50 months [32–37].

Calzavara et al. [33] investigated MAL-PDT for the treatment of BD and SCC, reporting an overall complete response rate of 87.8% at 3 months and 70.7% at 2 years in 41 biopsy-proven BD lesions [33]. In an observational, retrospective study, Truchuelo et al. [34] analyzed 51 BD tumors treated with MAL-PDT. The complete remission rate was 76.1%, and the recurrence rate was 14.3% after 1 year. A Swedish monocentric retrospective study on 423 BD lesions treated with MAL-PDT over a 13-year period found a complete response rate of 77.5% at the first follow-up visit (mean: 3.5 months) with a recurrence rate of 18.3% at the later follow-up visit (mean FU duration of 11.2 months, range 0.2–151 months) and an overall clearance rate of 63.4% [38]. The complete remission rates of small lesions (diameter < 20 mm) and large lesions (diameter > 20 mm) were 69.1% and 48.7%, respectively. A diameter greater than 20 mm was the main cause of treatment failure [38]. The other potential risk factors, i.e., gender, age, anatomic site, weeks between PDT sessions, and pain did not significantly correlate with the MAL-PDT efficacy rate.

Long-term MAL-PDT follow-up data have been reported in difficult-to-treat BD by Cavicchini and colleagues [35]. An analysis of 43 BD lesions showed a 100% complete response rate at 3 months and 89.4% at 50 months of follow-up. Jansen et al. [37] retrospectively studied 241 BD tumors treated with ALA or MAL-PDT (two sessions, one week apart) and found that the recurrence rate of BD tumors after 1 year and 5 years of PDT was 13.4% and 22.3%. In a large Spanish retrospective analysis of 537 BD lesions treated with MAL-PDT during the period 2006–2017, the 1-year and 5-year recurrence-free survival rates were 87% and 71%, respectively. Tumor size > 300 mm^2 ($\geq$21 mm in diameter), location on the upper extremities, and patient's age <70 years were all associated with an increased risk of recurrence [39].

Clinical, histological, and immunohistochemical variables implicated in the response to MAL-PDT were analyzed in a retrospective study including 33 BD lesions [40]. A response to MAL-PDT was observed in 82% of the lesions after 3 months of follow-up, decreasing to 70% after 6 years of follow-up. The tumor size was significantly larger in nonresponders than in responders (25 $\pm$ 8.7 mm vs. 14.9 $\pm$ 7.6 mm). No histological variables were associated with the response to MAL-PDT. P53 immunostaining was positive in a higher proportion of responders as compared to non-responders, while cyclin D1 and EGFR immunostainings were more intense in non-responders. On a multivariate analysis, p53 was the only variable that significantly correlated with response to MAL-PDT, with a possible role in increasing PpIX levels and subsequent cell death [40].

Real-world experiences with ALA-PDT have been reported in small studies, describing a complete response rate of 80–100% in the short-term and a relapse rate of about 0–10% at 12 months [41–43]. This efficacy rate is consistent with either one or two treatments, with 10% or 20% ALA and using the single or two-fold illumination scheme. In a long-term follow-up study including 19 BD lesions treated with a single session of 20% ALA-PDT, 89.5% achieved complete clearance at 3 months, with 76.5% still clear at 2 years, but only 53.3% at 5 years [44].

A retrospective study including 68 BD lesions treated with 20% ALA solution and blue light, with variable incubation time and total number of PDT treatments, reported an initial complete response rate of 77.9% within 3 months after the completion of all PDT treatments, which was not associated with the number of PDT treatments. On multivariate analysis, a longer ALA incubation time, smaller tumor diameter (<2 cm), and location on the face were all associated with increased effectiveness of PDT [45].

Inconclusive data have been published regarding the comparison between ALA and MAL for PDT for BD. One study comparing the efficacy of ALA and MAL in 27 BD lesions found a complete response rate of 89% with ALA and 78% with MAL at approximately

6 months with no significant difference [46]. In a large study including 191 BD lesions, complete response was obtained in 84.7% of the lesions after ALA-PDT and in 55.1% after MAL-PDT ($p < 0.001$) at the 12-month follow-up [47].

Table 1 summarizes relevant studies investigating PDT in the treatment of BD. None of the published studies separately analyzed the efficacy of PDT in treating BD lesions in sun-exposed and non-sun-exposed areas in terms of clinical response, recurrence rate, and cosmetic outcome. Figure 2 shows remission of BD after two sessions of MAL-PDT from our real life experience.

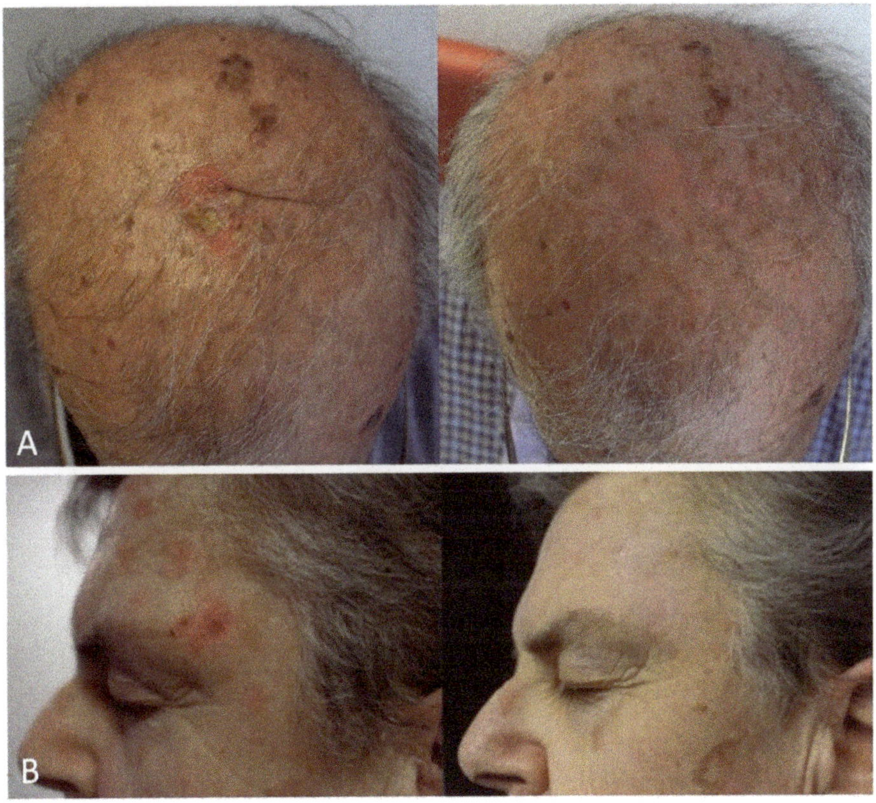

Figure 2. Treatment of Bowen's disease with MAL-PDT. (**A**) BD lesion on the scalp in a 76-year-old OTR patient before and after two sessions of MAL-PDT, 1 week apart; (**B**) A 64-year-old female patient with a BD tumor on the temporal region before and after MAL-PDT treatment.

Table 1. Relevant results from recent and/or larger studies on PDT for BD.

First Author, Year	No. of Patients	No. of Bowen's Disease Lesions	Protocol	Clinical Response Rate (%)	Follow-up (Months)	Recurrence Rate	Cosmetic Outcome
Dragieva G, 2004 [a] [48]	4 (4)	4	20% ALA emulsion PDT, visible incoherent light, 75 J/cm^2, 1 or 2 sessions	CR 100% at 1 month	12	50% at 12 months	Excellent in 100%
Morton C, 2006 [32]	96	111	MAL-PDT, red light 570–670 nm, 75 J/cm^2, 2 sessions 1 week apart	CR 73% after two sessions and CR 93% after four sessions, at 3 months	12	15% at 12 months	Good or excellent in 97%
Perrett CM, 2007 [a] [49]	8 (8)	9	MAL-PDT, red light 633 ± 15 nm, 37 J/cm^2, 2 sessions 1 week apart	CR 89% at 1 month	6	0 at 6 months	Excellent in 100%
De Haas ER, 2007 [42]	40	50	20% ALA ointment PDT, 1-fold illumination with diode 630 nm or LED light, 75 J/cm^2, or 2-fold illumination with LED light, 20 and 80 J/cm^2	CR 80%, single illumination CR 88%, 2-fold illumination at 12 months	24 (mean)	NA	Good in 92%
Calzavara-Pinton PG, 2008 [33]	NA	41	MAL-PDT, LED 635 nm, 37 J/cm^2, 2 sessions 1 week apart	CR 87.8% at 3 months	24	17.1% at 24 months	Excellent in 62%
Souza CS, 2009 [44]	19	19	20% ALA emulsion PDT, 630 nm diode laser, 100 and 300 J/cm^2, 1 session	CR 89.5% at 3 months	60	46.6% at 60 months	Good or excellent in 100%
Cavicchini S, 2011 [35]	30	43	MAL-PDT, red light 635 nm, 75 J/cm^2, 2 sessions 1 week apart	CR 100% at 6 months	50 (mean)	11.6% at 12 months	Excellent in 100%
Truchuelo M, 2012 [34]	42	46	MAL-PDT, red light 630 nm, 38 J/cm^2, 2 sessions 1 week apart	CR 76.1% PR 23.9%	16.6 (mean)	14.3% at 16.6 months (mean)	Excellent in 100%
Tarstedt M, 2016 [46]	NA	27	MAL-PDT or 20% ALA-PDT, red light 630 nm, 37 J/cm^2, 1 or 2 sessions few weeks apart	CR 78% MAL-PDT CR 89% ALA-PDT, at 6 months	6	NA	NA
Ratour-Bigot C, 2016 [a] [50]	105 (25)	151	MAL-PDT, red light 570–670 nm, 37 J/cm^2, 1 to 6 sessions	CR 52%, PR 26% At 3 months	14 (median)	NA	NA

Table 1. *Cont.*

First Author, Year	No. of Patients	No. of Bowen's Disease Lesions	Protocol	Clinical Response Rate (%)	Follow-up (Months)	Recurrence Rate	Cosmetic Outcome
Zaar, O 2017 [38]	335	423	ALA or MAL-PDT, red light 630 nm, 37–40 J/cm^2, 2 sessions 1 week apart	CR 77.5% at 3.5 months (mean)	11.2 (mean)	18.3% at 11.2 months (mean)	Excellent in 78%
Jansen MHE, 2018 [37]	241	NA	ALA or MAL-PDT, red light 630 nm, 37–40 J/cm^2, 2 sessions 1 week apart	NA	60	13.4% after 12 months; 22.3% after 60 months	NA
Gracia-Cazaña T, 2018 [40]	NA	33	MAL-PDT, red light 635 nm, 37 J/cm^2, 2 sessions 1 week apart	CR 82% at 3 months	72	12% at 72 months	NA
Aguilar-Bernier M, 2019 [39]	NA	537	MAL-PDT, red light 630 nm, 37 J/cm^2, 2 sessions 1 week apart	CR 88% at 12 months CR 71% at 60 months	33.2 (mean)	NA	NA
Alique-Garcìa S, 2019 [47]	171	191	MAL-PDT or 10% ALA gel, red light 635 nm, 37 J/cm^2, 1 or 2 cycle of two sessions 12 weeks apart	CR 76.5% MAL-PDT CR 87.3% ALA-PDT, at 3 months	12	27.88% MAL-PDT 1.8% ALA-PDT at 12 months	NA
Safar R, 2019 [51]	1	1	Daylight MAL-PDT, 1 session	Complete remission	3	NA	Good
Kibbi N, 2020 [45]	58	68	20% ALA solution PDT, non-coherent blue light, 400–500 nm, 10 J/cm^2, 1–4 sessions	CR 77.9% at 3 months	9.7 (median)	13.2% at 11.2 months (median)	NA
Martins CC, 2020 [52]	19	24	Daylight MAL-PDT, 2 sessions 1 week apart	CR 25%, PR 57% at 3 months	3	NA	NA
Cervantes JA, 2021 [43]	12	12	10% ALA gel PDT, 630 nm red light, 37 J/cm^2, 1 or 2 sessions 10 days apart	CR 100% at 1 month	1	NA	Good or excellent in 75.3%
González-Guerra E, 2023 [a] [41]	1 (1)	2	BF-200 ALA gel PDT, red light 630 nm, 37 J/cm^2, 3 sessions	PR 100% after treatment	12	100% at 12 months	NA
Ahmady S, 2024 [53]	78	78	MAL-PDT, red light 630 nm, 37 J/cm^2, 2 sessions 1 week apart	CR 82.1% at 12 months	12	NA	NA

Table 1. *Cont.*

First Author, Year	No. of Patients	No. of Bowen's Disease Lesions	Protocol	Clinical Response Rate (%)	Follow-up (Months)	Recurrence Rate	Cosmetic Outcome
COMBINED TREATMENTS							
Nakano A, 2011 [54]	4	4	20% ALA solution PDT, excimer-pumped dye laser radiation,630 nm, 50 J/cm^2 + Radiotherapy (3 Gy)	CR 100% at 3 months	14 (mean)	0 in 14 months (mean)	NA
Ko DY, 2013 [55]	21	58	MAL-PDT red light 632 nm, 37 J/cm^2 + Er:YAG ablative fractional laser, one session or MAL-PDT red light 632 nm, 37 J/cm^2, 2 sessions 1 week apart	CR 93.8% Er:YAG AFL-MAL-PDT CR 73.1% MAL-PDT at 3 months	12	6.7% Er:YAG AFL-MAL-PDT 31.6% MAL-PDT at 12 months	Excellent or good in 90.6% in the Er:YAG AFL-PDT group and 92.3% in the MAL-PDT group
Lu Y, 2014 [56]	13	13	Surgery + 10% ALA emulsion PDT, laser light 635 nm, 120 J/cm^2, 3 sessions	CR 100% at 6 months	12	0 at 12 months	NA
Cai H, 2015 [57]	10	11	20% ALA emulsion PDT, red light 630 nm, 180 J/cm^2 + CO_2 laser 2–3 W, 1–3 sessions	CR 72.7%, PR 27.3% at 1 month	6	9.1% at 6 months	NA
Victoria-Martìnez AM, 2017 [58]	10	13	MAL-PDT or 10% ALA nano-emulsion PDT, red light 632 nm, 37 J/cm^2, 3 sessions 1 week apart + Imiquimod 5% cream	CR 84.6%, PR 15.4% [b] at 3 months	18	18.1% at 18 months [b]	Very good in 100%
Genouw, E 2018 [59]	6	6	MAL-PDT, red light 635 nm, 37 J/cm^2 + CL (12 W) or FL (30 W) CO_2 laser, 2 sessions 2 weeks apart	CR 80%, PR 20% at 12 months	12	NA	Good or excellent in 100%
Wu Y, 2018 [60]	24	38	PBN-ALA-PDT or ALA-PDT, 10% ALA cream, LED, 633 nm, 100–200 J/cm^2	CR 77.8% PBN-ALA-PDT CR 40% ALA-PDT, at 1.5 months	12	0 PBN-ALA-PDT 11.7% ALA-PDT at 6 months	NA

Table 1. *Cont.*

First Author, Year	No. of Patients	No. of Bowen's Disease Lesions	Protocol	Clinical Response Rate (%)	Follow-up (Months)	Recurrence Rate	Cosmetic Outcome
Kim HJ, 2018 [61]	60	84	MAL-PDT red light 632 nm, 37 J/cm^2 + Er:YAG ablative fractional laser, one session or MAL-PDT red light 632 nm, 37 J/cm^2, 2 sessions 1 week apart	CR 93.48% Er:YAG AFL-MAL-PDT CR 76.3% MAL-PDT, at 3 months	60	9.3% Er:YAG AFL-MAL-PDT 41.38% MAL-PDT at 60 months	NA
Liu D, 2019 [62]	10	44	Simple shaving + 20% ALA cream PDT, 633 nm red light, 3 sessions 1 week apart	CR 100% at 3 months	12 (minimum)	0 at 12 months	Excellent in 100%
Liu X, 2023 [63]	11	12	Electrodesiccation + 20% ALA cream PDT, YAG LED-IB light, 633 nm ± 10 nm, 3 sessions	CR 100% at 12 months	17.5 (mean)	0 at 12 months	NA

ALA, aminolevulinic acid; *PDT*, photodynamic therapy; *MAL*, methyl aminolevulinate; *NA*, not available; *CR*, complete response, *PR*, partial response; *LED*, light emitting diode; *Er:YAG laser*, erbium-doped yttrium aluminium garnet laser; CO_2 *laser*, carbon dioxide laser; *CL*, continuous ablative laser; *FL*, fractional ablative laser; *AFL*, ablative fractional laser; *PBN*, plum-blossom needling. [a] The study included immunosuppressed patients. The number of immunocompromised patients is reported in the brackets. [b] Patients with partial response or recurrence were treated with topical 5% imiquimod.

A systematic review including nine studies assessed the different therapies for BD: PDT and 5-FU appeared effective, but due to limited evidence no clear conclusions on comparative efficacy were made. Surgical excision was not included because of the lack of comparative studies [36]. Later, Jansen et al. [37] retrospectively investigated the clinical efficacy of MAL or ALA-PDT, 5% 5-FU compared with surgical excision in 841 BD tumors. BD treated with 5-FU and PDT had a more than 2-fold increased 5-year probability of treatment failure compared with surgical excision, whereas there was no statistically significant difference between 5-FU and PDT. Of all treated BD, only eight tumors (seven post PDT and one post 5-FU) progressed into an invasive SCC, 3–42 months post-treatment. The same authors recently published a multicenter noninferiority trial comparing the effectiveness of 5% 5-FU cream twice daily for 4 weeks, 2 sessions of MAL-PDT with 1 week interval, and surgical excision in 250 patients with BD [53]. The proportion of patients with sustained clearance at 12 months was 97.4% after excision, 85.7% after 5-FU, and 82.1% after MAL-PDT. Based on the predefined noninferiority margin of 22%, 5-FU was noninferior to excision but associated with a better cosmetic outcome. For MAL-PDT, noninferiority to excision could not be concluded.

In a systematic review and meta-analysis including 12 randomized controlled trials published from 1996 to 2018, a higher lesion reduction rate after the first PDT treatment session was observed (OR = 2.86, 95%CI 1.89–4.33; $p < 0.00001$), with a significant difference versus both 5-FU (OR = 3.70; 95%CI: 2.07–6.62; $p < 0.00001$) and cryotherapy (OR = 2.24, 95%CI: 1.24–4.04; $p = 0.008$) [64]. However, no significant differences emerged in recurrence rates following treatment with PDT vs 5-FU (OR = 0.69; 95%CI 0.28–1.69) or cryotherapy (OR 0.53; 95%CI: 0.24–1.16). A more recently published meta-analysis including eight randomized controlled trials confirmed that PDT results in a significantly higher complete response rate (RR = 1.36, $p = 0.04$), reduced recurrences (RR = 0.53, $p = 0.03$), and better cosmetic outcomes (RR = 1.34, $p = 0.0002$) compared with other treatments, i.e., 5-FU and cryotherapy [65].

Evidence is very limited for daylight PDT (dlPDT). Two case reports of BD treated with dlPDT showed complete response in three BD lesions [51,66]. In a prospective study including 24 BD lesions treated with one cycle of 2 MAL-dlPDT sessions, complete clinical response was reported in 25% of the lesions, partial response in 57%, and no response in 16% of the lesions [52].

4. Combined Treatments

PDT has been combined with an ablative fractional resurfacing laser [55,61], CO_2 laser [57], electrodessication [63], surgery [56], radiation [54], imiquimod [58], plum-blossom needle [60], and simple shaving [62] for BD treatment.

Combination therapy with laser-assisted techniques has been consistently demonstrated to effectively increase the penetration depth of the photosensitizer as well as increase PDT's therapeutic effect. A small pilot randomized study supported the promising role of laser-assisted MAL-PDT in six BD lesions [59]. An efficacy of 80% was demonstrated with both continuous and fractional ablative CO_2-assisted MAL-PDT after 12 months. PDT illumination was significantly less painful in the fractional-assisted MAL-PDT group. Ko et al. [55] compared the efficacy, recurrence rate, cosmetic outcome, and safety between a single treatment with Er:YAG ablative fractional laser-assisted PDT (AFL)-PDT and standard MAL-PDT (two treatment sessions with a 1-week interval) in 58 BD lesions. At 12 months, Er:YAG AFL-PDT was more effective than MAL-PDT (93.8% vs. 73.1%; $p = 0.031$) and the recurrence rate was significantly lower for Er:YAG AFL-PDT than MAL-PDT (6.7% vs. 31.6%; $p = 0.022$). No difference was found in terms of cosmetic outcome or safety [55]. A long-term follow-up study investigated the 5-year efficacy and recurrence rates of AFL-MAL-PDT and conventional MAL-PDT for the treatment of BD on the lower extremities in 84 lesions [61]. After 5 years, the overall clearance rate of AFL-MAL-PDT was significantly better than that of MAL-PDT (84.78% vs. 44.74%, $p < 0.001$). The recurrence rate was significantly lower for AFL-MAL-PDT than for MAL-PDT (9.3% vs. 41.38%,

$p = 0.003$). Independent factors for treatment failure were a diameter larger than 20 mm and lesions previously treated.

Treatment with ALA-PDT combined with CO_2 laser was compared to CO_2 laser alone in a trial including 22 BD lesions. There was no difference in the complete remission rate (72.73% vs. 63.63%, $p > 0.05$); however, the recurrence rate at 6 months was significantly higher in the CO_2 laser alone than in the ALA-PDT plus CO_2 laser group (9.1% vs. 45.45%, $p < 0.05$) [57].

5. Non-Invasive Monitoring of Therapeutic Response

Dermoscopy can be used as a valuable follow-up tool in cases where non-surgical therapeutic options are chosen for the management of BD [17,67,68]. An illustrative example from our clinical experience is shown in Figure 3. Dermoscopic monitoring was performed 3 months after treatment in 23 patients with 29 histopathologically diagnosed BD lesions treated with MAL-PDT or imiquimod 5% cream [69]. Histopathological results showed that the cure rate for BD was 60% (3/5) for imiquimod cream and 50% (12/24) for MAL-PDT. After treatment, dermoscopic examination revealed the disappearance of pre-existing vascular structures in 16 lesions and residual vascular structures in 13 lesions. Histopathologic examination showed remnant intraepithelial neoplasias and increased vascularity in the dermis in lesions with persistent dermoscopic vascular structures. However, lesions without dermoscopic vascular structures showed normal epidermis and decreased vascularity in the dermis in all but one. During follow-up, one lesion showed a reappearance of vascular structures 9 months after treatment, which was confirmed to be a recurrence of BD after histopathology. These results supported the indications that emerged from a previous study by the same group [70].

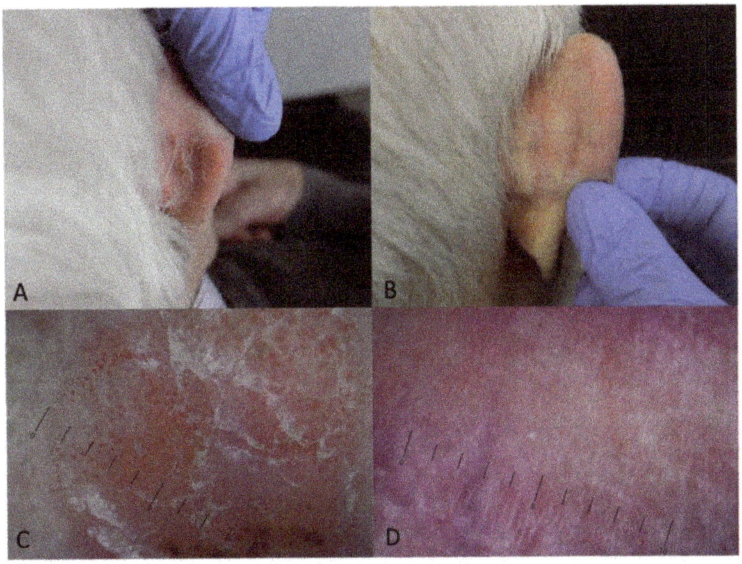

Figure 3. Dermoscopic monitoring (10×) of treatment response. Bowen's disease in an 89-year-old patient on the retro-auricular area before (**A–C**) and after two sessions of MAL-PDT (**B**,**D**).

RCM was useful to monitor for residual BD as well as to detect recurrence after PDT treatment in a case series including 10 patients with a total of 11 biopsies [71]. RCM imaging was found to help decrease unnecessary biopsies, especially in BD lesions that developed post-inflammatory erythema, which may make clinical and dermoscopic assessment difficult. In a case report, an in situ glans SCC was treated with two sessions of PDT using a copper bromide laser as a light source and the efficacy of the treatment was monitored with RCM [72]. After two sessions of PDT, RCM showed a normal mucosa, confirming the remission of the tumor.

6. Immunocompromised Patients

Immunocompromised BD patients appear to be significantly younger, more likely to have multiple tumors, and are at higher risk for recurrence and progression to invasive disease as compared to patients with normal immune function [27]. In addition, BD lesions occur multifocally and arise in body areas protected from UV light such as the trunk or the anogenital area.

Four BD lesions in transplant patients were treated with 1–2 sessions of ALA-PDT resulting in a complete response at 4 weeks; however, two patients experienced recurrence at 12 weeks [48]. One patient with two BD lesions underwent PDT with BF-200 ALA gel and red-light. The response (defined as an over 75% clearance of the lesion) was very good; however, incomplete resolution led to the recurrence of both lesions one year after treatment [73]. A randomized intrapatient comparative study found MAL-PDT more effective than 5% 5-FU in achieving a complete resolution of BD and AK lesions in eight OTR patients [49]. At 3-month follow up, the complete response rate for PDT was 89% (95% CI: 0.52–0.99), whilst for 5-FU it was 11% (95% CI: 0.003–0.48). At 6 months after treatment, the efficacy remained unchanged for both treatment groups. Unfortunately, the reported data do not allow discrimination between the response rates of BD and AK lesions.

Regarding the potential role of PDT in promoting the occurrence of SCC, a monocentric retrospective study investigated 105 patients with BD, including 25 (24%) immunocompromised patients, treated with MAL-PDT, who received a total of 151 different PDT fields. The efficacy of MAL-PDT was not significantly different between immunocompromised and non-immunocompromised patients. A total of 16 out of 105 patients developed SCC in PDT areas, after a median time of 6.0 months (IQR 2.7–11.8). The risk of the occurrence of at least one SCC in a PDT field was not significantly different between immunocompromised and non-immunocompromised patients [50].

Overall, limited evidence is available on the use of PDT for BD as well as on the comparison of PDT with other therapies in immunocompromised patients, making it difficult to draw conclusions; thus, treated patients should be closely monitored.

7. Tolerability and Cosmetic Outcome

Pain and burning during illumination, which peak in the first few minutes of treatment, are the main side effects of PDT. Expected skin phototoxicity effects are erythema, edema, vesiculation/pustulation, crusting, and erosion/ulceration. Long-term adverse effects such as pigmentary change, scarring or contact allergy, are uncommon. Systemic adverse events possibly related to the treatment have been very rarely reported [74].

Morton et al. [6] reported that most treatment-related events with MAL-PDT were considered as mild (60%) or moderate (34%), and only 6% were severe. By comparison, 12% of local events with cryotherapy were severe [6]. A higher severity of pain or burning during treatment, and of erythema after treatment, were observed in the MAL-PDT group compared to both excision and 5-FU ($p < 0.001$) [53]. When comparing MAL-PDT with ALA-PDT, no significant differences were identified in terms of high pain score (VAS, 8–10) (9% vs. 7%, respectively) and other frequent adverse events, such as erythema (41.9% vs. 43.6%), desquamation (37.5% vs. 32.7%), and superficial wounds (14% vs. 10.9%) [64].

In the study by Zaar et al. [38], the majority of BD lesions treated with MAL-PDT (195/250, 78.0%) healed with no long-term adverse events observed during follow-up. The most common adverse event was scarring, which was observed in 8.8% of the cases. Other local skin reactions were erythema (6%), hypopigmentation (2.4%), and hyperpigmentation (2.0%). Combinations of adverse events were seen in seven cases (2.8%).

The cosmetic outcome of MAL-PDT compares favorably with cryotherapy and 5% 5-FU in the treatment of BD lesions. At 3 months, MAL-PDT was superior to either cryotherapy or 5% 5-FU, with a good or excellent cosmetic outcome in 94% of patients vs.

66% for cryotherapy and 76% for 5% 5-FU, and was maintained at 12 months [6]. In addition, investigators and patients reported good/excellent outcomes significantly more often after MAL-PDT treatment than after excision ($p < 0.001$ and $p = 0.006$, respectively) [53].

8. Conclusions and Future Directions

PDT is a safe and effective, well-established treatment option for BD, especially in difficult locations, large or multiple lesions, and elderly patients. Lesion response appears to be significantly correlated with lesion size. Combination therapy with laser-assisted techniques has been shown to further improve PDT effectiveness. The published evidence on PDT both as monotherapy and combination therapy does not allow for adequate comparison because the protocols are different and the results regarding complete response and recurrence rate are variably reported. Noninvasive diagnostic techniques can help in the early diagnosis and treatment monitoring of BD. However, scientific evidence regarding treatment monitoring for BD is very limited, as it mainly focuses on actinic keratosis and basal cell carcinoma. Side effects, especially pain, are common, but generally mild, easily controlled, and self-limiting; patient information enables optimal management. The cosmetic outcome of MAL-PDT compares favorably with cryotherapy and 5% 5-FU with high levels of patient satisfaction. Patients with BD treated with PDT should be monitored after treatment because of the risk of incomplete response and recurrence, as well as progression to invasive cSCC, particularly for immunocompromised patients.

Considering the progressive aging of the general population, as well as the increase in immunosuppressed subjects, the incidence of both BD and cSCC is steadily increasing, constituting a growing public health problem. Future research on PDT for BD should focus on standardizing treatment protocols, improving the use of combination treatments, and encouraging studies of noninvasive methods in treatment monitoring, including those more recently introduced. Finally, for better patient selection, it would be desirable to promote large studies to identify additional predictors of clinical response and disease progression, beyond lesion size and immunosuppressive condition.

Author Contributions: Conceptualization, P.A. and M.C.F.; methodology, C.P. and M.C.F.; validation, P.A., C.P. and M.C.F.; formal analysis, P.A. and M.C.F.; investigation, P.A., C.P., C.C., M.B., L.D., M.M. and M.E.; data curation and search, P.A., C.P., C.C., M.B., L.D., M.M. and M.E.; writing—original draft preparation, P.A.; writing—review and editing, P.A., C.P., M.E. and M.C.F.; supervision, M.C.F.; project administration, M.C.F. and C.P. All authors have read and agreed to the published version of the manuscript.

Funding: This research received no external funding.

Conflicts of Interest: M.C.F. has served on advisory boards, received honoraria for lectures and/or research grants from AMGEN, Almirall, Abbvie, Boehringer-Ingelheim, BMS, Galderma, Kyowa Kyrin, Leo Pharma, Pierre Fabre, UCB, Lilly, Pfizer, Janssen, MSD, Novartis, Sanofi-Regeneron, and Sunpharma. M.E. has served as a speaker/board member for Abbvie, Almirall, Biogen, Celgene, Eli Lilly, Janssen, Leo Pharma, and Novartis. The remaining authors declare that the research was conducted in the absence of any commercial or financial relationships that could be construed as a potential conflict of interest. The funders had no role in the design of the study; in the collection, analyses, or interpretation of data; in the writing of the manuscript, or in the decision to publish the results.

References

1. Bowen, J.T. Precancerous dermatoses: A study of two cases of chronic atypical epithelial proliferation. *J. Cutan. Dis.* **1912**, *30*, 241–255. [CrossRef]
2. Callen, J.P. Bowen's disease and internal malignant disease. *Arch. Dermatol.* **1988**, *124*, 675–676. [CrossRef] [PubMed]
3. Arlette, J.P.; Trotter, M.J. Squamous cell carcinoma in situ of the skin: History, presentation, biology and treatment. *Australas. J. Dermatol.* **2004**, *45*, 1–11. [CrossRef] [PubMed]
4. Reizner, G.T.; Chuang, T.Y.; Elpern, D.J.; Stone, J.L.; Farmer, E.R. Bowen's disease (squamous cell carcinoma in situ) in Kauai, Hawaii. A population-based incidence report. *J. Am. Acad. Dermatol.* **1994**, *31*, 596–600. [CrossRef] [PubMed]

5. Moloney, F.J.; Comber, H.; O'Lorcain, P.; O'Kelly, P.; Conlon, P.J.; Murphy, G.M. A population-based study of skin cancer incidence and prevalence in renal transplant recipients. *Br. J. Dermatol.* **2006**, *154*, 498–504. [CrossRef] [PubMed]
6. Morton, C.A.; Birnie, A.J.; Eedy, D.J. British Association of Dermatologists' guidelines for the management of squamous cell carcinoma in situ (Bowen's disease) 2014. *Br. J. Dermatol.* **2014**, *170*, 245–260. [CrossRef] [PubMed]
7. Kim, J.Y.S.; Kozlow, J.H.; Mittal, B.; Moyer, J.; Olenecki, T.; Rodgers, P. Guidelines of care for the management of cutaneous squamous cell carcinoma. *J. Am. Acad. Dermatol.* **2018**, *78*, 560–578. [CrossRef] [PubMed]
8. Stratigos, A.J.; Garbe, C.; Dessinioti, C.; Lebbe, C.; Bataille, V.; Bastholt, L.; Dreno, B.; Fargnoli, M.C.; Forsea, A.M.; Frenard, C.; et al. European interdisciplinary guideline on invasive squamous cell carcinoma of the skin: Part 1. epidemiology, diagnostics and prevention. *Eur. J. Cancer* **2020**, *128*, 60–82. [CrossRef] [PubMed]
9. Miki, Y.; Kawatsu, T.; Matsuda, K.; Machino, H.; Kubo, K. Cutaneous and pulmonary cancers associated with Bowen's disease. *J. Am. Acad. Dermatol.* **1982**, *6*, 26–31. [CrossRef]
10. Watson, K.; Creamer, D. Arsenic-induced keratoses and Bowen's disease. *Clin. Exp. Dermatol.* **2004**, *29*, 46–48. [CrossRef]
11. Meyer, T.; Arndt, R.; Christophers, E.; Nindl, I.; Stockfleth, E. Importance of human papillomaviruses for the development of skin cancer. *Cancer Detect. Prev.* **2001**, *25*, 533–547. [PubMed]
12. Palaniappan, V.; Karthikeyan, K. Bowen's Disease. *Indian Dermatol. Online J.* **2022**, *13*, 177–189. [CrossRef] [PubMed]
13. Kossard, K.; Rosen, R. Cutaneous Bowen's disease. *J. Am. Acad. Dermatol.* **1992**, *27*, 406–410. [CrossRef] [PubMed]
14. Cox, N.H. Body site distribution of Bowen's disease. *Br. J. Dermatol.* **1994**, *130*, 714–716. [CrossRef] [PubMed]
15. Hansen, J.P.; Drake, A.L.; Walling, H.W. Bowen's Disease: A four-year retrospective review of epidemiology and treatment at a university center. *Dermatol. Surg.* **2008**, *34*, 878–883. [CrossRef] [PubMed]
16. Zalaudek, I.; Argenziano, G.; Leinweber, B.; Citarella, L.; Hofmann-Wellenhof, R.; Malvehy, J.; Puig, S.; Pizzichetta, M.A.; Thomas, L.; Soyer, H.P.; et al. Dermoscopy of Bowen's disease. *Br. J. Dermatol.* **2004**, *150*, 1112–1116. [CrossRef]
17. Lallas, A.; Argenziano, G.; Zendri, E.; Moscarella, E.; Longo, C.; Grenzi, L.; Pellacani, G.; Zalaudek, I. Update on non-melanoma skin cancer and the value of dermoscopy in its diagnosis and treatment monitoring. *Expert Rev. Anticancer. Ther.* **2013**, *13*, 541–558. [CrossRef] [PubMed]
18. Ianoși, S.L.; Batani, A.; Ilie, M.A.; Tampa, M.; Georgescu, S.R.; Zurac, S.; Boda, D.; Ianosi, N.G.; Neagoe, D.; Calina, D.; et al. Non-invasive imaging techniques for the in vivo diagnosis of Bowen's disease: Three case reports. *Oncol. Lett.* **2019**, *17*, 4094–4101. [CrossRef] [PubMed]
19. Ulrich, M.; Kanitakis, J.; González, S.; Lange-Asschenfeldt, S.; Stockfleth, E.; Roewert-Huber, J. Evaluation of Bowen disease by in vivo reflectance confocal microscopy. *Br. J. Dermatol.* **2012**, *166*, 451–453. [CrossRef]
20. Shahriari, N.; Grant-Kels, J.M.; Rabinovitz, H.S.; Oliviero, M.; Scope, A. Reflectance Confocal Microscopy Criteria of Pigmented Squamous Cell Carcinoma In Situ. *Am. J. Dermatopathol.* **2018**, *40*, 173–179. [CrossRef]
21. Cinotti, E.; Bertello, M.; Cartocci, A.; Fiorani, D.; Tognetti, L.; Solmi, V.; Cappilli, S.; Peris, K.; Perrot, J.L.; Suppa, M.; et al. Comparison of reflectance confocal microscopy and line-field optical coherence tomography for the identification of keratinocyte skin tumours. *Ski. Res. Technol.* **2023**, *29*, e13215. [CrossRef] [PubMed]
22. Bhawan, J. Squamous cell carcinoma in situ in skin: What does it mean? *J. Cutan. Pathol.* **2007**, *34*, 953–955. [CrossRef] [PubMed]
23. Granata, S.; Tessari, G.; Stallone, G.; Zaza, G. Skin cancer in solid organ transplant recipients: Still an open problem. *Front. Med.* **2023**, *10*, 1189680. [CrossRef]
24. Ulrich, C.; Arnold, R.; Frei, U.; Hetzer, R.; Neuhaus, P.; Stockfleth, E. Skin changes following organ transplantation: An interdisciplinary challenge. *Dtsch. Arztebl. Int.* **2014**, *111*, 188–194.
25. Garrett, G.L.; Blanc, P.D.; Boscardin, J.; Lloyd, A.A.; Ahmed, R.L.; Anthony, T.; Bibee, K.; Breithaupt, A.; Cannon, J.; Chen, A.; et al. Incidence of and Risk Factors for Skin Cancer in Organ Transplant Recipients in the United States. *JAMA Dermatol.* **2017**, *153*, 296–303. [CrossRef] [PubMed]
26. Fania, L.; Didona, D.; Di Pietro, F.R.; Verkhovskaia, S.; Morese, R.; Paolino, G.; Donati, M.; Ricci, F.; Coco, V.; Ricci, F.; et al. Cutaneous Squamous Cell Carcinoma: From Pathophysiology to Novel Therapeutic Approaches. *Biomedicines* **2021**, *9*, 171. [CrossRef]
27. Drake, A.L.; Walling, H.W. Variations in presentation of squamous cell carcinoma in situ (Bowen's disease) in immunocompromised patients. *J. Am. Acad. Dermatol.* **2008**, *59*, 68–71. [CrossRef]
28. Eimpunth, S.; Goldenberg, A.; Hamman, M.S.; Oganesyan, G.; Lee, R.A.; Hunnangkul, S.; Song, S.S.; Greywal, T.; Jiang, S.I.B. Squamous Cell Carcinoma In Situ Upstaged to Invasive Squamous Cell Carcinoma: A 5-Year, Single Institution Retrospective Review. *Dermatol. Surg.* **2017**, *43*, 698–703. [CrossRef]
29. Henderson, B.W.; Dougherty, T.J. How does photodynamic therapy work? *Photochem Photobiol.* **1992**, *55*, 145–157. [CrossRef]
30. Calzavara-Pinton, P.G.; Venturini, M.; Sala, R. Photodynamic therapy: Update 2006. Part 1: Photochemistry and photobiology. *J. Eur. Acad. Dermatol. Venereol.* **2007**, *21*, 293–302. [CrossRef]
31. Morton, C.A.; Szeimies, R.M.; Basset-Seguin, N.; Calzavara-Pinton, P.; Gilaberte, Y.; Haedersdal, M.; Hofbauer, G.F.L.; Hunger, R.E.; Karrer, S.; Piaserico, S.; et al. European Dermatology Forum guidelines on topical photodynamic therapy 2019 Part 1: Treatment delivery and established indications—Actinic keratoses, Bowen's disease and basal cell carcinomas. *J. Eur. Acad Dermatol. Venereol.* **2019**, *33*, 2225–2238. [CrossRef] [PubMed]

32. Morton, C.; Horn, M.; Leman, J.; Tack, B.; Bedane, C.; Tjioe, M.; Ibbotson, S.; Khemis, A.; Wolf, P. Comparison of topical methyl aminolevulinate photodynamic therapy with cryotherapy or Fluorouracil for treatment of squamous cell carcinoma in situ: Results of a multicenter randomized trial. *Arch. Dermatol.* **2006**, *142*, 729–735. [CrossRef] [PubMed]
33. Calzavara-Pinton, P.G.; Venturini, M.; Sala, R.; Capezzera, R.; Parrinello, G.; Specchia, C.; Zane, C. Methylaminolaevulinate-based photodynamic therapy of Bowen's disease and squamous cell carcinoma. *Br. J. Dermatol.* **2008**, *159*, 137–144. [CrossRef] [PubMed]
34. Truchuelo, M.; Fernandez-Guarino, M.; Fleta, B.; Alcántara, J.; Jaén, P. Effectiveness of photodynamic therapy in Bowen's disease: An observational and descriptive study in 51 lesions. *J. Eur. Acad. Dermatol. Venereol.* **2012**, *26*, 868–874. [CrossRef] [PubMed]
35. Cavicchini, S.; Serini, S.M.; Fiorani, R.; Girgenti, V.; Ghislanzoni, M.; Sala, F. Long-term follow-up of methyl aminolevulinate (MAL)-PDT in difficult-to-treat cutaneous Bowen's disease. *Int. J. Dermatol.* **2011**, *50*, 1002–1005. [CrossRef] [PubMed]
36. Bath-Hextall, F.J.; Matin, R.N.; Wilkinson, D.; Leonardi-Bee, J. Interventions for cutaneous Bowen's disease. *Cochrane Database Syst. Rev.* **2013**, CD007281. [CrossRef]
37. Jansen, M.H.; Appelen, D.; Nelemans, P.J.; Winnepenninckx, V.J.; Kelleners-Smeets, N.W.J.; Mosterd, K. Bowen's Disease: Long-term Results of Treatment with 5-Fluorouracil Cream, Photodynamic Therapy or Surgical Excision. *Acta Derm. Venereol.* 2018; 98, 114–115.
38. Zaar, O.; Fougelberg, J.; Hermansson, A.; Gillstedt, M.; Wennberg-Larkö, A.M.; Paoli, J. Effectiveness of photodynamic therapy in Bowen's disease: A retrospective observational study in 423 lesions. *J. Eur. Acad. Dermatol. Venereol.* **2017**, *31*, 1289–1294. [CrossRef] [PubMed]
39. Aguilar-Bernier, M.; Rodríguez-Barón, D.; Rivas-Ruiz, F.; Segura-Palacios, J.M.; de Troya Martín, M. Long-term efficacy of photodynamic therapy with methyl aminolevulinate in treating Bowen's disease in clinical practice: A retrospective cohort study (2006–2017). *Photodermatol. Photoimmunol. Photomed.* **2019**, *35*, 208–213. [CrossRef]
40. Gracia-Cazaña, T.; Salazar, N.; Vera-Álvarez, J.; Aguilera, J.; López-Navarro, N.; Herrera-Ceballos, E.; González, S.; Juarranz, Á.; Gilaberte, Y. Clinical, histological and immunohistochemical markers of resistance to methyl aminolaevulinate photodynamic therapy in Bowen disease. *Br. J. Dermatol.* **2018**, *178*, e138–e140. [CrossRef]
41. Cox, N.H.; Eedy, D.J.; Morton, C.A. Therapy Guidelines and Audit Subcommittee; British Association of Dermatologists. Guidelines for management of Bowen's disease: 2006 update. *Br. J. Dermatol.* **2007**, *156*, 11–21. [CrossRef]
42. De Haas, E.R.; Sterenborg, H.J.; Neumann, H.A.; Robinson, D.J. Response of Bowen disease to ALA-PDT using a single and a 2-fold illumination scheme. *Arch. Dermatol.* **2007**, *143*, 264–265. [CrossRef]
43. Cervantes, J.A.; Zeitouni, N.C. Photodynamic therapy utilizing 10% ALA nano-emulsion gel and red-light for the treatment of squamous cell carcinoma in-situ on the trunk and extremities: Pilot study and literature update. *Photodiagn. Photodyn. Ther.* **2021**, *35*, 102358. [CrossRef]
44. Souza, C.S.; Felicio, L.B.; Ferreira, J.; Kurachi, C.; Bentley, M.; Tedesco, A.; Bagnato, V. Long-term follow-up of topical 5-aminolaevulinic acid photodynamic therapy diode laser single session for non-melanoma skin cancer. *Photodiagn. Photodyn. Ther.* **2009**, *6*, 207–213. [CrossRef]
45. Kibbi, N.; Zhang, Y.; Leffell, D.J.; Christensen, S.R. Photodynamic therapy for cutaneous squamous cell carcinoma in situ: Impact of anatomic location, tumor diameter, and incubation time on effectiveness. *J. Am. Acad. Dermatol.* **2020**, *82*, 1124–1130. [CrossRef]
46. Tarstedt, M.; Gillstedt, M.; Wennberg Larkö, A.M.; Paoli, J. Aminolevulinic acid and methyl aminolevulinate equally effective in topical photodynamic therapy for non-melanoma skin cancers. *J. Eur. Acad. Dermatol. Venereol.* **2016**, *30*, 420–423. [CrossRef]
47. Alique-García, S.; Alique, D.; Company-Quiroga, J.; Sánchez, A.; Núñez, A.H.; Borbujo, J. Treatment of Bowen's disease with photodynamic therapy. Observational study in 171 patients with 5-aminolaevulinic acid (BF-200 ALA) and methyl aminolaevulinate (MAL). *Photodiagn. Photodyn. Ther.* **2019**, *28*, 192–194. [CrossRef]
48. Dragieva, G.; Hafner, J.; Dummer, R.; Schmid-Grendelmeier, P.; Roos, M.; Prinz, B.M.; Burg, G.; Binswanger, U.; Kempf, W. Topical photodynamic therapy in the treatment of actinic keratoses and Bowen's disease in transplant recipients. *Transplantation* **2004**, *77*, 115–121. [CrossRef]
49. Perrett, C.M.; McGregor, J.M.; Warwick, J.; Karran, P.; Leigh, I.M.; Proby, C.M.; Harwood, C.A. Treatment of post-transplant premalignant skin disease: A randomized intrapatient comparative study of 5-fluorouracil cream and topical photodynamic therapy. *Br. J. Dermatol.* **2007**, *156*, 320–328. [CrossRef]
50. Ratour-Bigot, C.; Chemidling, M.; Montlahuc, C.; Abirached, G.; Madjlessi, N.; Bullier, C.; Battistella, M.; Bagot, M.; Lebbe, C.; Basset-Seguin, N. Squamous Cell Carcinoma Following Photodynamic Therapy for Cutaneous Bowen's Disease in a Series of 105 Patients. *Acta Derm. Venereol.* **2016**, *96*, 658–663. [CrossRef]
51. Safar, R.; Alkhars, A.; Tallegas, M.; Korsaga-Some, N.; Machet, L. Successful treatment for extensive Bowen's disease using daylight-mediated photodynamic therapy. *Acta Derm. Venereol.* **2019**, *99*, 701–702. [CrossRef]
52. Martins, C.C.; Bakos, R.M.; Martins Costa, M. Daylight photodynamic therapy for Bowen's disease. *An. Bras. Dermatol.* **2020**, *95*, 529–531. [CrossRef] [PubMed]
53. Ahmady, S.; Nelemans, P.J.; Kelleners-Smeets, N.W.J.; Arits, A.H.M.M.; de Rooij, M.J.M.; Kessels, J.P.H.M.; Essers, B.A.B.; Mosterd, K. Surgical excision versus topical 5% 5-fluorouracil and photodynamic therapy in treatment of Bowen's disease: A multicenter randomized controlled trial. *J. Am. Acad. Dermatol.* **2024**, *90*, 58–65. [CrossRef] [PubMed]
54. Nakano, A.; Watanabe, D.; Akita, Y.; Kawamura, T.; Tamada, Y.; Matsumoto, Y. Treatment efficiency of combining photodynamic therapy and ionizing radiation for Bowen's disease. *J. Eur. Acad. Dermatol. Venereol.* **2011**, *25*, 475–478. [CrossRef] [PubMed]

55. Ko, D.Y.; Kim, K.H.; Song, K.H. A randomized trial comparing methyl aminolaevulinate photodynamic therapy with and without Er:YAG ablative fractional laser treatment in Asian patients with lower extremity Bowen disease: Results from a 12-month follow-up. *Br. J. Dermatol.* **2014**, *170*, 165–172. [CrossRef] [PubMed]
56. Lu, Y.G.; Wang, Y.Y.; Yang, Y.D.; Zhang, X.C.; Gao, Y.; Yang, Y.; Zhang, J.B.; Li, G.L. Efficacy of topical ALA-PDT combined with excision in the treatment of skin malignant tumor. *Photodiagn. Photodyn. Ther.* **2014**, *11*, 122–126. [CrossRef] [PubMed]
57. Cai, H.; Wang, Y.X.; Zheng, J.C.; Sun, P.; Yang, Z.Y.; Li, Y.L.; Liu, X.Y.; Li, Q.; Liu, W. Photodynamic therapy in combination with CO_2 laser for the treatment of Bowen's disease. *Lasers Med. Sci.* **2015**, *30*, 1505–1510. [CrossRef]
58. Victoria-Martinez, A.M.; Martinez-Leborans, L.; Ortiz-Salvador, J.M.; Perez-Ferriols, A. Treatment of Bowen Disease with Photodynamic Therapy and the Advantages of Sequential Topical Imiquimod. *Actas Dermo-Sifiliogr.* **2017**, *108*, e9–e14. [CrossRef]
59. Genouw, E.; Verheire, B.; Ongenae, K.; De Schepper, S.; Creytens, D.; Verhaeghe, E.; Boone, B. Laser-assisted photodynamic therapy for superficial basal cell carcinoma and Bowen's disease: A randomized intrapatient comparison between a continuous and a fractional ablative CO2 laser mode. *J. Eur. Acad. Dermatol. Venereol.* **2018**, *32*, 1897–1905. [CrossRef] [PubMed]
60. Wu, Y.; Wang, P.; Zhang, L.; Wang, B.; Wang, X. Enhancement of Photodynamic Therapy for Bowen's Disease Using Plum-Blossom Needling to Augment Drug Delivery. *Dermatol. Surg.* **2018**, *44*, 1516–1524. [CrossRef]
61. Kim, H.J.; Song, K.H. Ablative fractional laser-assisted photodynamic therapy provides superior long-term efficacy compared with standard methyl aminolevulinate photodynamic therapy for lower extremity Bowen disease. *J. Am. Acad. Dermatol.* **2018**, *79*, 860–868. [CrossRef]
62. Liu, D.; Wu, L.; Li, J.; Shi, W.; Li, F.; Su, J.; Huang, K.; Zhou, Q.; Zhao, S.; Chen, M. Simple shaving combined with photodynamic therapy for refractory bowen disease. *Photodiagn. Photodyn. Ther.* **2019**, *26*, 258–260. [CrossRef] [PubMed]
63. Liu, X.; Wang, J.; Yu, J.; Xing, W.; Zhang, J. Experience analysis of a combined photodynamic/electrodesiccation therapy in the treatment of 11 cases of large patches of Bowen's disease. *Photodiagn. Photodyn. Ther.* **2023**, *43*, 103710. [CrossRef] [PubMed]
64. Zhong, S.; Zhang, R.; Mei, X.; Wang, L. Efficacy of photodynamic therapy for the treatment of Bowen's disease: An updated systematic review and meta-analysis of randomized controlled trials. *Photodiagn. Photodyn. Ther.* **2020**, *32*, 102037. [CrossRef] [PubMed]
65. Xue, W.L.; Ruan, J.Q.; Liu, H.Y.; He, H.X. Efficacy of Photodynamic Therapy for the Treatment of Bowen's Disease: A Meta-Analysis of Randomized Controlled Trials. *Dermatology* **2022**, *238*, 542–550. [CrossRef] [PubMed]
66. Pérez-Pérez, L.; García-Gavín, J.; Gilaberte, Y. Daylight-mediated photodynamic therapy in Spain: Advantages and disadvantages. *Actas Dermosifiliogr.* **2014**, *105*, 663–674. [CrossRef] [PubMed]
67. Fargnoli, M.C.; Kostaki, D.; Piccioni, A.; Micantonio, T.; Peris, K. Dermoscopy in the diagnosis and management of non-melanoma skin cancers. *Eur. J. Dermatol.* **2012**, *22*, 456–463. [CrossRef]
68. Guida, S.; Alma, A.; Shaniko, K.; Chester, J.; Ciardo, S.; Proietti, I.; Giuffrida, R.; Zalaudek, I.; Manfredini, M.; Longo, C.; et al. Non-Melanoma Skin Cancer Clearance after Medical Treatment Detected with Noninvasive Skin Imaging: A Systematic Review and Meta-Analysis. *Cancers* **2022**, *14*, 2836. [CrossRef]
69. Mun, J.H.; Park, J.M.; Song, M.; Jwa, S.W.; Kim, H.S.; Ko, H.C.; Kim, B.S.; Kim, M.B. The use of dermatoscopy to monitor therapeutic response of Bowen disease: A dermatoscopic pathological study. *Br. J. Dermatol.* **2012**, *167*, 1382–1385. [CrossRef]
70. Mun, J.H.; Kim, S.H.; Jung, D.S.; Ko, H.C.; Kwon, K.S.; Kim, M.B. Dermoscopic features of Bowen's disease in Asians. *J. Eur. Acad. Dermatol. Venereol.* **2010**, *24*, 805–810. [CrossRef]
71. Teoh, Y.L.; Kuan, L.Y.; Chong, W.S.; Chia, H.Y.; Thng, T.G.S.; Chuah, S.Y. The role of reflectance confocal microscopy in the diagnosis and management of squamous cell carcinoma in situ treated with photodynamic therapy. *Int. J. Dermatol.* **2019**, *58*, 1382–1387. [CrossRef]
72. Cinotti, E.; Perrot, J.L.; Labeille, B.; Douchet, C.; Mottet, N.; Cambazard, F. Laser photodynamic treatment for in situ squamous cell carcinoma of the glans monitored by reflectance confocal microscopy. *Australas. J. Dermatol.* **2014**, *55*, 72–74. [CrossRef] [PubMed]
73. González-Guerra, E.; Taboada, A.C.; Muñoz, L.C.; Fructuoso, A.I.S. Photodynamic therapy with BF-200 ALA gel for the treatment of actinic keratosis, Bowen's disease and basal cell carcinoma in long term immunosuppressed patients after organ transplantation. *Photodiagn. Photodyn. Ther.* **2023**, *45*, 103882. [CrossRef] [PubMed]
74. Ibbotson, S.H.; Wong, T.H.; Morton, C.A.; Collier, N.J.; Haylett, A.; McKenna, K.E.; Mallipeddi, R.; Moseley, H.; Rhodes, L.E.; Seukeran, D.C.; et al. Adverse effects of topical photodynamic therapy: A consensus review and approach to management. *Br. J. Dermatol.* **2019**, *180*, 715–729. [CrossRef] [PubMed]

MDPI AG
Grosspeteranlage 5
4052 Basel
Switzerland
Tel.: +41 61 683 77 34

Biomedicines Editorial Office
E-mail: biomedicines@mdpi.com
www.mdpi.com/journal/biomedicines

www.ingramcontent.com/pod-product-compliance
Lightning Source LLC
Chambersburg PA
CBHW062059220926
51850CB00010B/2822
* 9 7 8 3 7 2 5 8 4 9 7 1 0 *